Analysis of Pavement Structures

Analysis of Pavement Structures brings together current research and existing knowledge on the analysis and design of pavements and introduces load and thermal stress analyses of asphalt and concrete pavement structures in a simple and step-by-step manner. For the second edition of this book, a new chapter on numerical implementation (using FEM) of pavement analysis is added along with topics such as mechanical modeling of granular materials, applications of convolution theorems in visco-elasticity, visco-elastic Poisson's ratio, concepts of fracture mechanics in relation to fatigue of asphalt mix, solution of semi-infinite and so forth. New solved examples and schematic diagrams are also added.

Features:

- Presents a simple, step-by-step approach for pavement analysis including systematic compilation of research work in the area
- Discusses further elaborations in terms of extended analytical formulations on some selected topics
- Includes new chapter on finite element analysis for pavement structures
- Contains more solved examples to understand the concepts better
- Explores primary application of pavement analysis in pavement thickness design

This book is aimed at graduate students, structural mechanics researchers, and senior undergraduate students in civil/pavement/highway/transport engineering.

Analysis of Pavement Structures

Second Edition

Animesh Das
Professor, Department of
Civil Engineering,
Indian Institute of Technology Kanpur,
Kanpur, India

CRC Press
Taylor & Francis Group
Boca Raton London New York

CRC Press is an imprint of the
Taylor & Francis Group, an informa business

Second edition published 2023
by CRC Press
6000 Broken Sound Parkway NW, Suite 300, Boca Raton, FL 33487-2742

and by CRC Press
4 Park Square, Milton Park, Abingdon, Oxon, OX14 4RN

First edition published by CRC Press 2015

CRC Press is an imprint of Taylor & Francis Group, LLC

ISBN: 978-1-032-04156-8 (hbk)
ISBN: 978-1-032-04157-5 (pbk)
ISBN: 978-1-003-19076-9 (ebk)

DOI: 10.1201/9781003190769

Typeset in Nimbus font
by KnowledgeWorks Global Ltd.

Dedication

Gifted to my beloved wife, Nibedita

Contents

List of Figures

Preface to the second edition

The basic organization and the approach of the book in this second edition are the same as they were in the first publication. The book continues to maintain a simple approach towards developing formulations for analysis of pavement structure from the basics and then linking these with various other documents for further studies.

A few solved examples and new schematic diagrams have been added. A few new sections have been added at different chapter. A new chapter on applications of finite element method (FEM) analysis method has been added.

I am thankful to some of my students for the useful discussions, namely, Aditya Agarwal, Satyendra Kumar Patel, Praveen Kumar, Abhinash Dev Pandey and others. I am also thankful to Dr. Binanda Narzary (Tezpur University), Dr. Nikhil Saboo (IIT Roorkee), Dr.-Ing. J. Stefan Bald (TU-Darmstadt) and Dr. Amar Nath Roy Chowdhury (IIT Kanpur) for their comments and discussions.

Preface to the first publication

This is a simple book.

This book is about pavement analysis. A pavement is a mutli-layered structure made up of a number of layers placed one over the other. These layers can be made of asphaltic material, cement concrete, bound or unbound stone aggregates, etc. These materials show complex mechanical responses with the variation of stress, time or temperature. Thus, understanding the performance of an in-service pavement structure subjected to vehicular loading and environmental variations is a difficult task.

However, this is a simple book. This book presents a step-by-step formulation for analyses of load and thermal stresses of idealized pavement structures. Some of these idealizations involve assumptions of material being linear, elastic, homogeneous and isotropic; load being static; thermal profile being linear; and so on.

Significant research has been done on analysis and design of pavements during the last half-a-century (some of the fundamental developments are, however, more than a hundred years old), and a large number of research publications are already available. However, there is a need for the basic formulations to be systematically compiled and put together in one place – hence, this book. It is believed that such a compilation will provide an exposure to the basic approaches used in pavement analyses and subsequently help the readers formulate their own research or field problems – more difficult than those dealt with in this book.

The idea of this book originated when I initiated a post-graduate course on *Characterization of pavement materials and analysis of pavements* at IIT Kanpur. This course was introduced in 2007; that was the time we were revising the post-graduate course structure in our Department. I must thank my colleague Dr. Partha Chakroborty, for suggesting at that time that the Transportation Engineering programme at IIT Kanpur should have a Pavement Engineering course with more analysis content. I also should thank him for his constant encouragement during the entire process of preparation of this manuscript. One of our graduate students, Ms. Priyanka Khan, typed out portions of my lectures as class notes. Those initial pages helped me to overcome the inertia of getting started to write this book. That is how it began.

This book has eight chapters. The first chapter introduces the sign convention followed in the book and mentions some of the basic solid mechanics formulations used in the subsequent chapters. The second chapter deals with the material characterization of various pavement materials. It introduces simple rheological models for asphaltic material. Beams and plates on elastic foundation are dealt with in the third chapter – these formulations form the basis of analysis of concrete pavement slab due to load. The fourth chapter covers the thermal stress in concrete pavement, it provides formulation for axial and bending stresses due to full and partial restraint conditions. The fifth chapter starts with the analysis of elastic half-space and enlarges it to analysis of multi-layered structure. A formulation for thermo-rheological analysis of asphalt pavement is presented in the sixth chapter. The seventh chapter discusses the pavement design principles where pavement analyses results are used. Finally, the last chapter discusses some miscellaneous topics which includes analysis of beam/ plate resting on elastic half-space, analysis for dynamic loading conditions, analysis of composite pavement, reliability issues in pavement design and inverse problems in pavement engineering.

Since this book provides an overview of basic approaches for pavement analysis, rigorous derivations for complex situations have been deliberately skipped. However, references are provided at appropriate contexts for further reading for readers who want to explore further. I must place a disclaimer that those references are not necessarily the only and the best reading material, but they are just a few representative.

A number of my former and present students have contributed towards development of this book. They have asked me questions inside the classroom and outside. Discussions with some of them were quite useful, while others helped me to cross-check a few derivations. Dr. Pabitra Rajbongshi, Sudhir N. Varma, Vivek Agarwal, Pranamesh Chakraborty, Syed Abu Rehan and Vishal Katariya are some of these students. I must also thank my former colleague, Dr. Ashwini Kumar, for the useful discussions with him on plate theories. I must also thank all the people with whom I have interacted with professionally from time-to-time discussing various issues related to pavement engineering. I also thank all the authors of numerous papers, books and other documents whose works have been referred to in this compilation book.

I am grateful to my parents for their care and thoughts involved in my education. I thank my father, Dr. Kali Charan Das, for teaching me formulations in physics using first principles. I thank my Ph.D. supervisor, Dr. B. B. Pandey, for training me as a researcher in pavement engineering. I also wish to thank my colleagues for the encouragement and my Institute, IIT Kanpur, for providing an excellent academic ambiance.

I thank the CRC Press & Taylor and Francis staff members, especially Dr. Gagandeep Singh and Ms. Joselyn Banks-Kyle for remaining in touch with me during the development of the manuscript, and Karen Simon during the production process. I must thank my family members for bearing with me, and especially my daughter, Anwesha, for sacrificing her bedtime stories for months, while I was finalizing the manuscript.

Care has been taken as far as possible to check the editorial mistakes. If, however, you find any, or wish to provide your feedback on this book, please drop me an e-mail at adas@iitk.ac.in.

This book, now, must go for printing...

(Animesh Das)

Table 1

List of symbols - I

a	Radius of circular area or equivalent tyre imprint
A	Area
A_{st}	Cross-sectional area of steel per unit length
a_{st}	Cross-sectional area of a single steel bar
B	Width of a concrete slab
$[B]$	Strain–displacement relationship matrix
C_{crp}	Creep compliance
C^h	Heat capacity
$C(t)$	Pseudo-stiffness
$[C]$	Constitutive matrix
D_c	Diameter
D	Flexural rigidity of a plate
E	Young's modulus or elastic modulus
E'	Storage modulus
E''	Loss modulus
E^*	Complex modulus
E_d	Dynamic modulus
E_{rel}	Relaxation modulus
F	Force
f	Coefficient of friction
g	Acceleration due to gravity
G	Shear modulus
h	Layer thickness
I_1	First stress invariant ($= \sigma_1 + \sigma_2 + \sigma_3$)
I_2	Second stress invariant
I_3	Third stress invariant
K	Bulk modulus
k	Modulus of subgrade reaction
k_i	Regression constants/coefficients
k^{td}	Coefficient of thermal diffusivity
k_s	Spring constant

Table 2

List of symbols - II

k_{ss}	Slider spring constant
$[k]$	Elemental stiffness matrix
$[K]$	Global stiffness matrix
l	Radius of relative stiffness
L	Length of a concrete slab
M	Moment
M_R	Resilient modulus
M_c	Unit cost of maintenance
M_u	Unit road user cost
n	Number of repetitions applied
N_{lf}	Laboratory fatigue life
N	Number of traffic repetitions a material/pavement can sustain
$[N]$	Shape function
P_a	Atmospheric pressure
q	Pressure (distributed load)
Q	Concentrated load
Q_h	Heat flow per unit area
$[Q]$	Global force matrix
x	Distance along X direction
y	Distance along Y direction
z	Distance along Z direction
r	Discount rate
R	Reliability
$[R]$	Rotation matrix
$R(x)$	Residual function
S	Structural health of a pavement
t	Time
T	Temperature
T	Number of traffic expected repetitions
T_t	Temperature at the top surface
T_b	Temperature at the bottom surface
T_∞	Temperature at infinite depth

Table 3

List of symbols - III

u	Displacement along X direction (Cartesian coordinate)
u_r	Displacement along R direction (cylindrical coordinate)
U	Total potential energy of the deformed body
U_s	Strain energy
U_g	Universal gas constant
$\{u\}$	Displacement field
$\{u'\}$	Nodal displacement matrix of the element
$[U]$	Global displacement matrix
V	Shear force
V_o	Speed
v	Displacement along Y direction (Cartesian coordinate)
v_θ	Displacement along tangential direction (Cylindrical coordinate)
W_i	Dissipated energy of a linearly viscoelastic material in the ith cycle
W_T	Cumulative dissipated energy of linearly viscoelastic material
w	Displacement along Z direction
z_s	Gap between two adjacent concrete slabs
α	Coefficient of thermal expansion
α_T	Time shift factor
β	An angle
$\{q\}$	Nodal force matrix of the element
δ	Phase angle
ΔH	Apparent activation energy
ε	Strain
ε_R	Pseudo strain

Table 4

List of symbols - IV

γ	Engineering shear strain
ζ	Dummy variable for time
θ	An angle in cylindrical polar coordinate system
η_d	Viscosity of the dashpot
ν	Poisson's ratio
ρ	Density of the material
Ω	Volume
σ	Normal stress
σ^c	Confining pressure ($= \sigma_3$)
σ_d	Deviatoric stress
σ^S	Tensile strength
σ^{TB}	Bending stress component due to temperature
σ^{TA}	Axial stress component due to temperature
σ^{TN}	Nonlinear stress component due to temperature
ω_f	Angular frequency
τ^b	Bond strength

Table 5

List of abbreviations

BF	Body force
DER	Dissipated energy ratio
IDT	Indirect tensile strength
FEM	Finite element method

About the book

The features of this book include the following:

- This book is a systematic compilation of works done in the area of analysis of pavement structure. Uniformity in the symbols and sign conventions has been maintained throughout while consolidating this knowledge in this book.
- This book provides a simple, step-by-step approach for pavement analysis using numerous schematic diagrams.
- The fundamental concepts and the governing principles have been emphasized so that the readers gain knowledge and capability to independently approach any pavement analysis problem.
- This is a book where analyses of both concrete and asphalt pavement structure due to vehicular as well as thermal loading have been presented.
- A conceptual link has been presented in the book (as in Chapter 8) on how the analyses' results can be utilized in the structural design of pavements.
- The book provides a quick reference to other books/documents/papers for further discussions on specific topics.

About the author

Animesh Das, Ph.D., is presently working as a Professor in the Department of Civil Engineering, Indian Institute of Technology Kanpur. He received his Ph.D. degree from Indian Institute of Technology Kharagpur. Dr. Das's area of interest is pavement material characterization, pavement analysis, pavement design and pavement maintenance. He has many technical publications in various journals of repute and in conference proceedings. He has co-authored a textbook titled *Principles of Transportation Engineering* published by the Prentice-Hall of India (currently, PHI Learning). He has received a number of awards in recognition to his contribution to the field, including the Indian National Academy of Engineers Young Engineer award (2004) and Fulbright-Nehru Senior Research Fellowship (2012–13). Details of his works can be found on his webpage: http://home.iitk.ac.in/~adas.

1 Introduction

1.1 PURPOSE OF THE BOOK

Pavement is a multi-layered structure. It is generally made up of compacted soil, bound or unbound granular material (stone aggregates), asphalt mix or cement concrete put as horizontal layers one above the other. Figure 1.1 presents a typical cross section of an asphalt and a concrete pavement. A composite pavement should have one layer made up of asphalt mix and another layer made up of cement concrete or cemented material (bound granular material).

Generally, asphalt pavements do not have joints, whereas the concrete pavements (commonly known as jointed plain cement concrete pavement) have joints[1]. The concrete pavements are made up of concrete slabs of finite dimensions with connections (generally of steels bars) to the adjacent slabs. Dowel bars are provided along the transverse joint, and tie bars are provided along the longitudinal joint (refer Figure 1.3). The dowel bars participate in the load transfer when the wheel moves from one slab to the other, thereby reducing the differential deflection between the two slabs at that time. The tie bars keep the two adjacent slabs in position. The design requirement of steel per unit length of concrete slab (along the transverse or longitudinal direction as the case may be) is therefore higher for dowel bars than that of tie bars.

Based on the functions, the joints in concrete pavements can be broadly divided into expansion joints and contraction joints. This is shown schematically in Figure 1.2 as the longitudinal section[2]. Expansion joints (placed transversely) allow the slabs to expand or contract due to temperature variations. Dowel bars are typically placed in the expansion joints – one end of the dowel bar is kept free (within the slab) to allow expansion/contraction of the bar (refer Figure 1.2). Contraction joints (also placed transversely) are generally provided as partial depth pre-cracks (as grooves) to allow the concrete slab to contract due to natural shrinkage (refer Figure 1.2). Thus, with the passage of time, the width of these grooves may increase and the cracks may extend downward to the full depth of the slab.

A block pavement or a segmental pavement is made up of inter-connected blocks (generally, cement concrete blocks), and its structural behaviour is different than usual asphalt or concrete pavements.

An in-service pavement is continuously subjected to traffic loading and temperature variations. The purpose of this book is to present a conceptual framework on the basic formulation of load and thermal stresses of typical concrete and asphalt pavements (as shown in Figures 1.1 and 1.3).

[1]Though exceptions are possible, for example, asphalt pavements in extreme cold climate may as well be provided with joints, continuously reinforced concrete pavements do not generally have joints, etc.

[2]Joint filler and sealant are not shown in Figure 1.2.

DOI: 10.1201/9781003190769-1

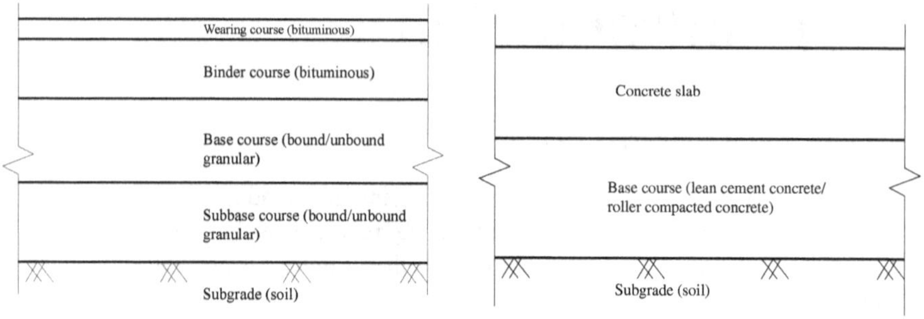

(a) A typical asphalt pavement section (b) A typical concrete pavement section

Figure 1.1 Typical cross section of an asphalt and a concrete pavement

The analysis of pavement structure enables one to predict or explain the pavement response to load from physical understanding of the governing principles, which can be corroborated later through experimental observations. This, in turn, builds confidence in structural design, evaluation and maintenance planning of road infrastructure.

Generally speaking, a concrete pavement is idealized as a plate resting on an elastic foundation [130, 317–319]. It is assumed that the load is transferred through bending, and the slab thickness does not undergo any change while it is subjected to load. For an asphalt pavement, it is assumed that the load is transferred through contacts of particles, and the layer thicknesses do undergo changes due to the application of load [41, 42, 146].

Pavement being a multi-layered structure, it is generally difficult to obtain a closed-form solution of its response due to load, because of the algebric complexities associated with the analytical formulations. For design purpose, pavement analysis may be done through some software which may use certain numerical methods to perform the analysis. Ready-to-use analysis charts are also available in various

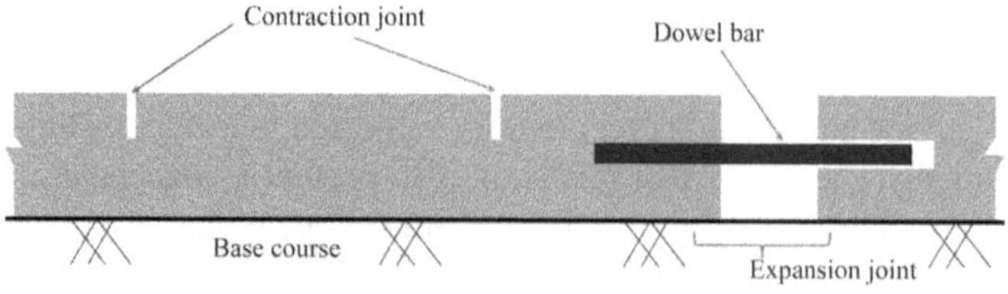

Figure 1.2 Longitudinal section of concrete pavement showing contraction and expansion joint

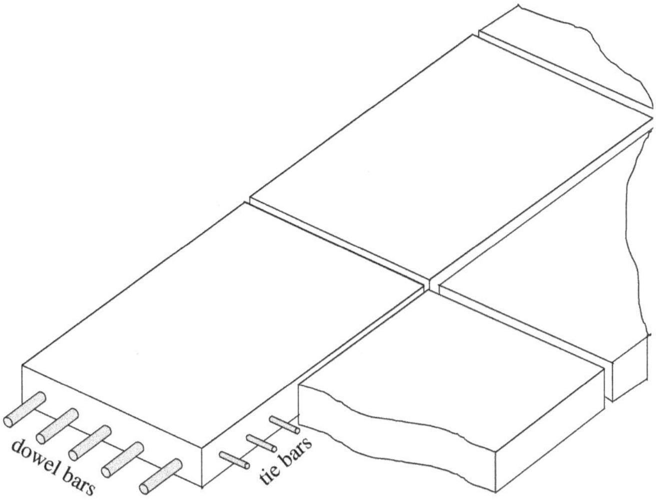

Figure 1.3 Dowel bar and tie bar arrangement in concrete pavement

codes/guidelines. The algorithms used and the assumptions involved in the analysis process may not always be apparent to a pavement designer (as user of these softwares/charts). Thus, there is a need to know the assumptions involved and the basic formulations needed for analysing a pavement structure.

On the other hand, large pool of research papers, books, reports, theses, etc., is available dealing with pavement material characterization and structural analyses of pavements. The research publications typically deal with a rigorous theoretical development on a specific aspect and may choose to skip details of some of the basic and well-known principles/techniques. Further, some of the basic formulations might have been developed many years ago (often with different notations and sign conventions than practiced today) and contributed by researchers from other areas of science and engineering. For instance, contributions to the theoretical formulation for pavement analysis have come from structural engineering (for example, theory of plates), soil mechanics (for example, beams and plates on elastic foundation), engineering mechanics (for example, principles of rheology), mechanical engineering (for example, material modelling), and so on.

For a beginner (in pavement engineering), this may appear as a hurdle – because, one may need to trace back the original source, understand the assumptions/idealizations and then follow the subsequent theoretical development. Thus, there is a need that the approaches to pavement analyses are collated at one place. This is the purpose of this book. The focus of the present book, therefore, can be identified as follows:

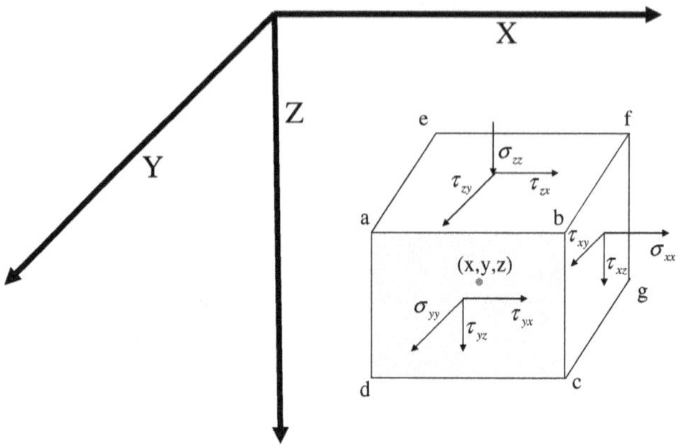

Figure 1.4 Sign convention in Cartesian system followed in this book

- This book is a compilation of existing knowledge on analyses of pavement structure. Load and thermal stress analyses for both asphalt and concrete pavements are dealt in this book. Attempts have been made to provide ready references to other publications/documents for further reading.
- Basic formulations for analysis of pavement structure have been presented in this book in a step-by-step manner – from a simple formulation to a more complex one. Attempts have been made to maintain a uniformity in symbol and sign conventions throughout the book.

1.2 BACKGROUND AND SIGN CONVENTIONS

Some of the basic and widely used equations, which are also referred and used in this book, are presented here almost without discussion. One can refer to books on mechanics of solids and theory of elasticity [15, 107, 249, 288] for further study; applications of some of these formulations can also be found in books on soil mechanics [70, 114, 169, 230]. The sign convention followed in the present book is shown in Figure 1.4 for Cartesian coordinate system and in Figure 1.5 for cylindrical coordinate system. These figures also show the notations used to identify the stresses in different directions.

1.2.1 STATE OF STRESS

The state of stress in Cartesian coordinate system (refer Figure 1.4) can be written as

$$[\sigma] = \begin{bmatrix} \sigma_{xx} & \tau_{xy} & \tau_{xz} \\ \tau_{yx} & \sigma_{yy} & \tau_{yz} \\ \tau_{zx} & \sigma_{zy} & \sigma_{zz} \end{bmatrix} \tag{1.1}$$

$[\sigma]$ is also known as Cauchy's stress.

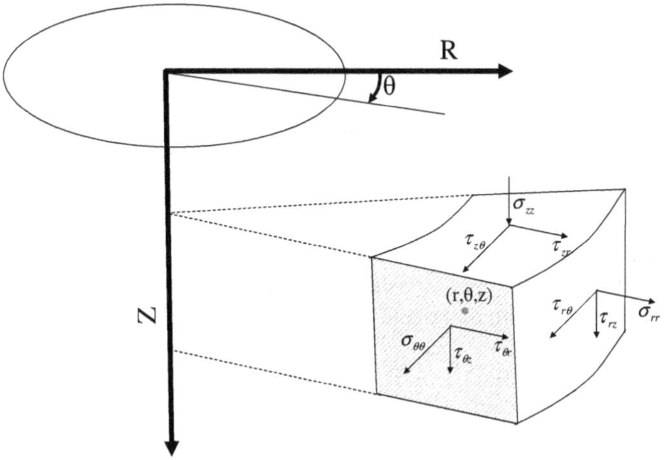

Figure 1.5 Sign convention in cylindrical system followed in this book

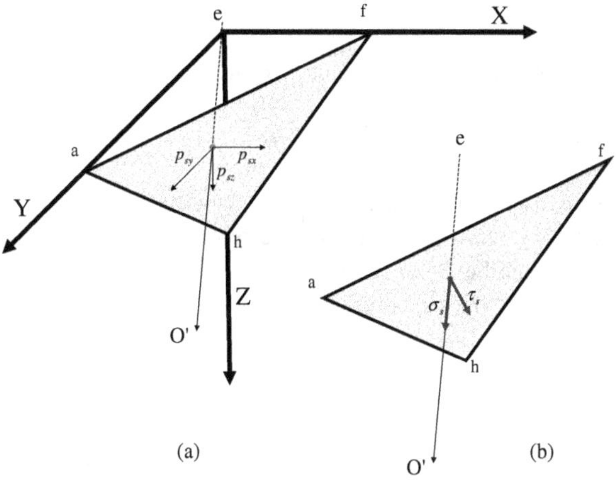

Figure 1.6 Stresses on plane 'afh' plane with direction cosines as l, m and n

Figure 1.6 shows a plane 'afh', whose direction cosine values are l, m and n with respect to X, Y and Z axes, respectively. If, p_{sx}, p_{sy} and p_{sz} represent the stresses (on plane 'afh') parallel to X, Y and Z (refer Figure 1.6(a)), then

$$\left\{ \begin{array}{c} p_{sx} \\ p_{sy} \\ p_{sz} \end{array} \right\} = \left[\begin{array}{ccc} \sigma_{xx} & \tau_{yx} & \tau_{zx} \\ \tau_{xy} & \sigma_{yy} & \tau_{zy} \\ \tau_{xz} & \sigma_{yz} & \sigma_{zz} \end{array} \right] \left\{ \begin{array}{c} l \\ m \\ n \end{array} \right\} \tag{1.2}$$

The normal stress in the plane 'afh' (refer Figure 1.6b) can be obtained as

$$\sigma_s = p_{sx}l + p_{sy}m + p_{sz}n \tag{1.3}$$

The shear stress on plane 'afh' is obtained as

$$\tau_s = \left(\left(p_{sx}^2 + p_{sy}^2 + p_{sz}^2\right) - \sigma_s^2\right)^{1/2} \tag{1.4}$$

For a special case, when the choice of the plane 'afh' (that is choice of l, m and n) is such that the shear stress vanishes, and therefore, only the normal stress exists (that is, principal stress on that plane), it can be written as

$$\begin{Bmatrix} \sigma_s l \\ \sigma_s m \\ \sigma_s n \end{Bmatrix} = \begin{bmatrix} \sigma_{xx} & \tau_{yx} & \tau_{zx} \\ \tau_{xy} & \sigma_{yy} & \tau_{zy} \\ \tau_{xz} & \sigma_{yz} & \sigma_{zz} \end{bmatrix} \begin{Bmatrix} l \\ m \\ n \end{Bmatrix} \tag{1.5}$$

For a solution to exist,

$$\begin{vmatrix} \sigma_{xx} - \sigma_s & \tau_{yx} & \tau_{zx} \\ \tau_{xy} & \sigma_{yy} - \sigma_s & \tau_{zy} \\ \tau_{xz} & \sigma_{yz} & \sigma_{zz} - \sigma_s \end{vmatrix} = 0 \tag{1.6}$$

The Equation 1.6 gives rise to the characteristic equation as

$$\sigma_s^3 - I_1\sigma_s^2 + I_2\sigma_s - I_3 = 0 \tag{1.7}$$

where I_1, I_2 and I_3 are coefficients. Since the principal stresses for a given state of stress should not vary with the choice of the reference coordinate axes, the coefficients I_1, I_2 and I_3 must have constant values. These are known as stress invariants. Their expressions are provided as Equations 1.8–1.10.

$$I_1 = \sigma_{xx} + \sigma_{yy} + \sigma_{zz} \tag{1.8}$$

$$I_2 = \sigma_{xx}\sigma_{yy} + \sigma_{yy}\sigma_{zz} + \sigma_{zz}\sigma_{xx} - \tau_{xy}^2 - \tau_{yz}^2 - \tau_{zx}^2 \tag{1.9}$$

$$I_3 = \sigma_{xx}\sigma_{yy}\sigma_{zz} + 2\tau_{xy}\tau_{yz}\tau_{zx} - \sigma_{xx}\tau_{yz}^2 - \sigma_{yy}\tau_{xz}^2 - \sigma_{zz}\tau_{xy}^2 \tag{1.10}$$

In terms of principal stresses, these take the following form

$$I_1 = \sigma_1 + \sigma_2 + \sigma_3 \tag{1.11}$$

$$I_2 = \sigma_1\sigma_2 + \sigma_2\sigma_3 + \sigma_3\sigma_1 \tag{1.12}$$

$$I_3 = \sigma_1\sigma_2\sigma_3 \tag{1.13}$$

There can be a special case when l, m, n values are all equal. That is, $l = m = n = \frac{1}{\sqrt{3}}$. The corresponding plane is known as octahedral plane. The octahedral normal stress (σ_{oct}) can be obtained from Equation 1.3 as follows:

$$\sigma_{oct} = \frac{1}{3}(\sigma_{xx} + \sigma_{yy} + \sigma_{zz}) \tag{1.14}$$

The octahedral shear stress (τ_{oct}) can be obtained from Equation 1.4 as follows:

$$\tau_{oct} = \frac{1}{3}\left((\sigma_{xx} - \sigma_{yy})^2 + (\sigma_{yy} - \sigma_{zz})^2 + (\sigma_{zz} - \sigma_{xx})^2 + 6\tau_{xy}^2 + 6\tau_{yz}^2 + 6\tau_{zx}^2\right)^{1/2} \tag{1.15}$$

The σ_{oct} and τ_{oct} in terms of principal stresses can be expressed as

$$\sigma_{oct} = \frac{1}{3}(\sigma_1 + \sigma_2 + \sigma_3) \tag{1.16}$$

$$\begin{aligned} \tau_{oct}^2 &= \frac{1}{9}\left((\sigma_1 - \sigma_2)^2 + (\sigma_2 - \sigma_3)^2 + (\sigma_3 - \sigma_1)^2\right) \\ &= \frac{2}{9}(I_1^2 - 3I_2) \end{aligned} \tag{1.17}$$

1.2.2 STRAIN – DISPLACEMENT AND STRAIN COMPATIBILITY EQUATIONS

If displacement fields in the direction of X, Y and Z are considered as u, v, and w, then under the assumptions of small deformation,

$$\begin{aligned} \varepsilon_{xx} &= \frac{\delta u}{\delta x} \\ \varepsilon_{yy} &= \frac{\delta v}{\delta y} \\ \varepsilon_{zz} &= \frac{\delta w}{\delta z} \\ \gamma_{xy} &= \frac{\delta u}{\delta y} + \frac{\delta v}{\delta x} \\ \gamma_{yz} &= \frac{\delta v}{\delta z} + \frac{\delta w}{\delta y} \\ \gamma_{zx} &= \frac{\delta w}{\delta x} + \frac{\delta u}{\delta z} \end{aligned} \tag{1.18}$$

where ε_{xx}, ε_{yy} and ε_{zz} indicate normal strains (along X, Y and Z directions), and γ_{xy}, γ_{yz} and γ_{zx} indicate the engineering shear strains.

In matrix form this can be written as

$$\{\varepsilon\} = [B]\{u\} \tag{1.19}$$

where $[B]$ can be called as strain–displacement relationship matrix.

By taking partial derivatives and suitably substituting, one can develop the following set of equations which do not contain the u, v and w terms. These equations (Equation set 1.20) are known as strain compatibility equations.

$$\begin{aligned}
\frac{\partial^2 \gamma_{xy}}{\partial x \partial y} &= \frac{\partial^2 \varepsilon_{xx}}{\partial y^2} + \frac{\partial^2 \varepsilon_{yy}}{\partial x^2} \\
\frac{\partial^2 \gamma_{yz}}{\partial y \partial z} &= \frac{\partial^2 \varepsilon_{yy}}{\partial z^2} + \frac{\partial^2 \varepsilon_{zz}}{\partial y^2} \\
\frac{\partial^2 \gamma_{zx}}{\partial z \partial x} &= \frac{\partial^2 \varepsilon_{zz}}{\partial x^2} + \frac{\partial^2 \varepsilon_{xx}}{\partial z^2} \\
2\frac{\partial^2 \varepsilon_{xx}}{\partial y \partial z} &= -\frac{\partial^2 \gamma_{yz}}{\partial x^2} + \frac{\partial^2 \gamma_{zx}}{\partial x \partial y} + \frac{\partial^2 \gamma_{xy}}{\partial x \partial z} \\
2\frac{\partial^2 \varepsilon_{yy}}{\partial z \partial x} &= -\frac{\partial^2 \gamma_{zx}}{\partial y^2} + \frac{\partial^2 \gamma_{xy}}{\partial y \partial z} + \frac{\partial^2 \gamma_{yz}}{\partial y \partial x} \\
2\frac{\partial^2 \varepsilon_{zz}}{\partial x \partial y} &= -\frac{\partial^2 \gamma_{xy}}{\partial z^2} + \frac{\partial^2 \gamma_{yz}}{\partial z \partial x} + \frac{\partial^2 \gamma_{zx}}{\partial z \partial y}
\end{aligned} \tag{1.20}$$

The strain–displacement relationships in cylindrical coordinate are

$$\begin{aligned}
\varepsilon_{rr} &= \frac{\partial u_r}{\partial r} \\
\varepsilon_{\theta\theta} &= \frac{u_r}{r} + \frac{1}{r}\frac{\partial v_\theta}{\partial \theta} \\
\varepsilon_{zz} &= \frac{\partial w}{\partial z} \\
\gamma_{r\theta} &= \frac{1}{r}\frac{\partial u_r}{\partial \theta} + \frac{\partial v_\theta}{\partial r} - \frac{v_\theta}{r} \\
\gamma_{rz} &= \frac{\partial u_r}{\partial z} + \frac{\partial w}{\partial r} \\
\gamma_{r\theta} &= \frac{\partial v_\theta}{\partial z} + \frac{1}{r}\frac{\partial w}{\partial \theta}
\end{aligned} \tag{1.21}$$

where $u_r =$ displacement in r direction, $v_\theta =$ displacement along tangential direction, $w =$ displacement along Z direction.

1.2.3 CONSTITUTIVE RELATIONSHIP BETWEEN STRESS AND STRAIN

The constitutive relationship for a linear anisotropic material can be written as

$$\begin{Bmatrix} \sigma_{xx} \\ \sigma_{yy} \\ \sigma_{zz} \\ \tau_{xy} \\ \tau_{yz} \\ \tau_{zx} \end{Bmatrix} = \begin{bmatrix} C_{11} & C_{12} & C_{13} & C_{14} & C_{15} & C_{16} \\ C_{21} & C_{22} & C_{23} & C_{24} & C_{25} & C_{26} \\ C_{31} & C_{32} & C_{33} & C_{34} & C_{35} & C_{36} \\ C_{41} & C_{42} & C_{43} & C_{44} & C_{45} & C_{46} \\ C_{51} & C_{52} & C_{53} & C_{54} & C_{55} & C_{56} \\ C_{61} & C_{62} & C_{63} & C_{64} & C_{65} & C_{66} \end{bmatrix} \begin{Bmatrix} \varepsilon_{xx} \\ \varepsilon_{yy} \\ \varepsilon_{zz} \\ \gamma_{xy} \\ \gamma_{yz} \\ \gamma_{zx} \end{Bmatrix} \tag{1.22}$$

where C_{ij} are the material constants. $[C]$ is also known as constitutive matrix.

Thus,

$$\{\sigma\} = [C]\{\varepsilon\} \tag{1.23}$$

For transversely isotropic material (that is, when the properties are same along a plane perpendicular to the axis of symmetry), Equation 1.22 takes the following form:

$$\begin{Bmatrix} \sigma_{xx} \\ \sigma_{yy} \\ \sigma_{zz} \\ \tau_{xy} \\ \tau_{yz} \\ \tau_{zx} \end{Bmatrix} = \begin{bmatrix} C_{11} & C_{12} & C_{13} & 0 & 0 & 0 \\ C_{12} & C_{11} & C_{13} & 0 & 0 & 0 \\ C_{13} & C_{13} & C_{33} & 0 & 0 & 0 \\ 0 & 0 & 0 & C_{44} & 0 & 0 \\ 0 & 0 & 0 & 0 & C_{44} & 0 \\ 0 & 0 & 0 & 0 & 0 & \frac{1}{2}(C_{11}-C_{12}) \end{bmatrix} \begin{Bmatrix} \varepsilon_{xx} \\ \varepsilon_{yy} \\ \varepsilon_{zz} \\ \gamma_{xy} \\ \gamma_{yz} \\ \gamma_{zx} \end{Bmatrix} \tag{1.24}$$

That means Equation 1.24 requires five independent constants to capture the constitutive relationship of transversely isotropic material.

For isotropic material (that is, when the properties are same along any direction), Equation 1.22 takes the following form:

$$\begin{Bmatrix} \sigma_{xx} \\ \sigma_{yy} \\ \sigma_{zz} \\ \tau_{xy} \\ \tau_{yz} \\ \tau_{zx} \end{Bmatrix} = \begin{bmatrix} C_{11} & C_{12} & C_{12} & 0 & 0 & 0 \\ C_{12} & C_{11} & C_{12} & 0 & 0 & 0 \\ C_{12} & C_{12} & C_{11} & 0 & 0 & 0 \\ 0 & 0 & 0 & \frac{1}{2}(C_{11}-C_{12}) & 0 & 0 \\ 0 & 0 & 0 & 0 & \frac{1}{2}(C_{11}-C_{12}) & 0 \\ 0 & 0 & 0 & 0 & 0 & \frac{1}{2}(C_{11}-C_{12}) \end{bmatrix} \begin{Bmatrix} \varepsilon_{xx} \\ \varepsilon_{yy} \\ \varepsilon_{zz} \\ \gamma_{xy} \\ \gamma_{yz} \\ \gamma_{zx} \end{Bmatrix} \tag{1.25}$$

where $C_{11} = \frac{E(1-\nu)}{(1+\nu)(1-2\nu)}$ and $C_{12} = \frac{E\nu}{(1+\nu)(1-2\nu)}$, E is the Young's modulus of the material. Thus, Equation 1.25 can be written as

$$\begin{Bmatrix} \sigma_{xx} \\ \sigma_{yy} \\ \sigma_{zz} \\ \tau_{xy} \\ \tau_{yz} \\ \tau_{zx} \end{Bmatrix} = \frac{E}{(1+\nu)(1-2\nu)} \begin{bmatrix} 1-\nu & \nu & \nu & 0 & 0 & 0 \\ \nu & 1-\nu & \nu & 0 & 0 & 0 \\ \nu & \nu & 1-\nu & 0 & 0 & 0 \\ 0 & 0 & 0 & \frac{1-2\nu}{2} & 0 & 0 \\ 0 & 0 & 0 & 0 & \frac{1-2\nu}{2} & 0 \\ 0 & 0 & 0 & 0 & 0 & \frac{1-2\nu}{2} \end{bmatrix} \begin{Bmatrix} \varepsilon_{xx} \\ \varepsilon_{yy} \\ \varepsilon_{zz} \\ \gamma_{xy} \\ \gamma_{yz} \\ \gamma_{zx} \end{Bmatrix} \tag{1.26}$$

The inverted form of Equation 1.26 can be presented as the following:

$$\begin{Bmatrix} \varepsilon_{xx} \\ \varepsilon_{yy} \\ \varepsilon_{zz} \\ \gamma_{xy} \\ \gamma_{yz} \\ \gamma_{zx} \end{Bmatrix} = \frac{1}{E} \begin{bmatrix} 1 & -\nu & -\nu & 0 & 0 & 0 \\ -\nu & 1 & -\nu & 0 & 0 & 0 \\ -\nu & -\nu & 1 & 0 & 0 & 0 \\ 0 & 0 & 0 & 2(1+\nu) & 0 & 0 \\ 0 & 0 & 0 & 0 & 2(1+\nu) & 0 \\ 0 & 0 & 0 & 0 & 0 & 2(1+\nu) \end{bmatrix} \begin{Bmatrix} \sigma_{xx} \\ \sigma_{yy} \\ \sigma_{zz} \\ \tau_{xy} \\ \tau_{yz} \\ \tau_{zx} \end{Bmatrix} \tag{1.27}$$

Equation 1.27 can be written as

$$\begin{aligned} \varepsilon_{xx} &= \frac{1}{E}(\sigma_{xx} - \nu(\sigma_{yy} + \sigma_{zz})) \\ \varepsilon_{yy} &= \frac{1}{E}(\sigma_{yy} - \nu(\sigma_{zz} + \sigma_{xx})) \\ \varepsilon_{zz} &= \frac{1}{E}(\sigma_{zz} - \nu(\sigma_{xx} + \sigma_{yy})) \end{aligned} \tag{1.28}$$

$$\begin{aligned}\gamma_{xy} &= \frac{1}{G}\tau_{xy}\\ \gamma_{yz} &= \frac{1}{G}\tau_{yz}\\ \gamma_{zx} &= \frac{1}{G}\tau_{zx}\end{aligned} \tag{1.29}$$

where $G = \frac{E}{2(1+\nu)}$. In cylindrical coordinate (for isotropic material), the equations are

$$\begin{aligned}\sigma_{rr} &= \frac{E}{(1+\nu)(1-2\nu)}((1-\nu)\varepsilon_{rr}+\nu\varepsilon_{zz}+\nu\varepsilon_{\theta\theta})\\ \sigma_{zz} &= \frac{E}{(1+\nu)(1-2\nu)}(\nu\varepsilon_{rr}+(1-\nu)\varepsilon_{zz}+\nu\varepsilon_{\theta\theta})\\ \sigma_{\theta\theta} &= \frac{E}{(1+\nu)(1-2\nu)}(\nu\varepsilon_{rr}+\nu\varepsilon_{zz}+(1-\nu)\varepsilon_{\theta\theta})\end{aligned} \tag{1.30}$$

$$\begin{aligned}\tau_{rz} &= \frac{E}{2(1+\nu)}\gamma_{rz}\\ \tau_{r\theta} &= \frac{E}{2(1+\nu)}\gamma_{r\theta}\\ \tau_{z\theta} &= \frac{E}{2(1+\nu)}\gamma_{z\theta}\end{aligned} \tag{1.31}$$

1.2.3.1 Plane Stress Condition (in Cartesian Coordinate System)

For plane stress condition, stress in one particular direction (say along the Y-axis direction) is zero. That is, $\sigma_{yy} = \tau_{xy} = \tau_{yz} = 0$. Such situation arises, for example, for a disk with a negligible thickness. Putting these conditions in Equations 1.28 and 1.29, one can write

$$\begin{aligned}\varepsilon_{xx} &= \frac{1}{E}(\sigma_{xx}-\nu\sigma_{zz})\\ \varepsilon_{yy} &= \frac{1}{E}(-\nu\sigma_{zz}-\nu\sigma_{xx})\\ \varepsilon_{zz} &= \frac{1}{E}(\sigma_{zz}-\nu\sigma_{xx})\end{aligned} \tag{1.32}$$

$$\begin{aligned}\gamma_{xy} &= 0\\ \gamma_{yz} &= 0\\ \gamma_{zx} &= \frac{1}{G}\tau_{zx} = \frac{2(1+\nu)}{E}\tau_{zx}\end{aligned} \tag{1.33}$$

Conversely, the stresses (in plane stress case) can be expressed (considering, $G = \frac{E}{2(1+\nu)}$) in terms of strains as follows:

$$\begin{aligned}
\sigma_{xx} &= \frac{E}{1-\nu^2}\varepsilon_{xx} + \frac{\nu E}{1-\nu^2}\varepsilon_{zz} \\
\sigma_{zz} &= \frac{E}{1-\nu^2}\varepsilon_{zz} + \frac{\nu E}{1-\nu^2}\varepsilon_{xx} \\
\tau_{zx} &= \frac{E}{2(1+\nu)}\gamma_{zx}
\end{aligned} \tag{1.34}$$

1.2.3.2 Plane Strain Condition (in Cartesian Coordinate System)

In plane strain condition, strain in one particular direction (say along Y direction) is zero. That is, $\varepsilon_{yy} = \gamma_{xy} = \gamma_{yz} = 0$. Such situation arises, for example, for an embankment, which has negligible strain along the longitudinal direction. Putting these conditions in Equations 1.28 and 1.29, one can write

$$\begin{aligned}
\varepsilon_{xx} &= \frac{1-\nu^2}{E}\sigma_{xx} - \frac{\nu(1+\nu)}{E}\sigma_{zz} \\
\varepsilon_{zz} &= \frac{1-\nu^2}{E}\sigma_{zz} - \frac{\nu(1+\nu)}{E}\sigma_{xx} \\
\gamma_{zx} &= \frac{2(1+\nu)}{E}\tau_{zx}
\end{aligned} \tag{1.35}$$

Conversely, the stresses (in plane strain case) can be expressed in terms of strains as follows:

$$\begin{aligned}
\sigma_{xx} &= \frac{E(1-\nu)}{(1+\nu)(1-2\nu)}\varepsilon_{xx} + \frac{\nu E}{(1+\nu)(1-2\nu)}\varepsilon_{zz} \\
\sigma_{zz} &= \frac{\nu E}{(1+\nu)(1-2\nu)}\varepsilon_{xx} + \frac{(1-\nu)E}{(1+\nu)(1-2\nu)}\varepsilon_{zz} \\
\tau_{zx} &= \frac{E}{2(1+\nu)}\gamma_{zx}
\end{aligned} \tag{1.36}$$

1.2.3.3 Axi-Symmetric Condition

For axi-symmetric situation (that is, when the material property, geometry and loading are symmetric about the axis of revolution), the response of the material/medium will be independent of θ, that is, $\frac{\partial}{\partial\theta} = 0$ and $v_\theta = 0$ (refer Equation 1.21) and hence $\tau_{r\theta} = \tau_{z\theta} = 0$. Putting these conditions in Equations 1.30 and 1.31, one can write

$$\begin{aligned}
\sigma_{rr} &= \frac{E}{(1+\nu)(1-2\nu)}((1-\nu)\varepsilon_{rr} + \nu\varepsilon_{zz} + \nu\varepsilon_{\theta\theta}) \\
\sigma_{zz} &= \frac{E}{(1+\nu)(1-2\nu)}(\nu\varepsilon_{rr} + (1-\nu)\varepsilon_{zz} + \nu\varepsilon_{\theta\theta}) \\
\sigma_{\theta\theta} &= \frac{E}{(1+\nu)(1-2\nu)}(\nu\varepsilon_{rr} + \nu\varepsilon_{zz} + (1-\nu)\varepsilon_{\theta\theta})
\end{aligned} \tag{1.37}$$

$$\tau_{rz} = \frac{E}{2(1+\nu)}\gamma_{rz} \tag{1.38}$$

1.2.4 EQUILIBRIUM EQUATIONS

The equilibrium condition is derived by taking force balance along each direction. The static equilibrium equation (in Cartesian coordinate system) can be written as

$$\begin{aligned}\frac{\partial \sigma_{xx}}{\partial x}+\frac{\partial \tau_{yx}}{\partial y}+\frac{\partial \tau_{zx}}{\partial z}+BF_x &= 0\\ \frac{\partial \tau_{xy}}{\partial x}+\frac{\partial \sigma_{yy}}{\partial y}+\frac{\partial \tau_{zy}}{\partial z}+BF_y &= 0\\ \frac{\partial \tau_{xz}}{\partial x}+\frac{\partial \tau_{yz}}{\partial y}+\frac{\partial \sigma_{zz}}{\partial z}+BF_z &= 0\end{aligned} \tag{1.39}$$

where BF_x, BF_y and BF_z are the body forces per unit volume (say, gravity, magnetic force, etc., as applicable) along X, Y and Z directions, respectively. For dynamic equilibrium case, the right-hand side of the equations will have terms as $\rho\frac{\partial^2 u}{\partial t^2}$, $\rho\frac{\partial^2 v}{\partial t^2}$ and $\rho\frac{\partial^2 w}{\partial t^2}$, respectively, where ρ is density of the material. The static equilibrium condition in cylindrical coordinate system is obtained as

$$\begin{aligned}\frac{\partial \sigma_{rr}}{\partial r}+\frac{1}{r}\frac{\partial \tau_{r\theta}}{\partial \theta}+\frac{\partial \tau_{zr}}{\partial z}+\frac{\sigma_{rr}-\sigma_{\theta\theta}}{r}+BF_r=0\\ \frac{\partial \tau_{r\theta}}{\partial r}+\frac{1}{r}\frac{\partial \sigma_{\theta\theta}}{\partial \theta}+\frac{\partial \tau_{\theta z}}{\partial z}+\frac{2\tau_{r\theta}}{r}+BF_\theta=0\\ \frac{\partial \tau_{zr}}{\partial r}+\frac{1}{r}\frac{\partial \tau_{\theta z}}{\partial \theta}+\frac{\partial \sigma_{zz}}{\partial z}+\frac{\tau_{zr}}{r}+BF_z=0\end{aligned} \tag{1.40}$$

where BF_r, BF_θ and BF_z are the body forces per unit volume along r, θ and Z directions, respectively.

1.3 CLOSURE

Some basic relationships in mechanics of solids and the theory of elasticity are recapitulated in this chapter as a background material. These equations have been referred/used at various places in the subsequent chapters of this book. One can refer to any suitable book on mechanics of solids and theory of elasticity for further discussions on these.

2 Material Characterization

2.1 INTRODUCTION

As mentioned in Section 1.1, different materials are used to build a road. This chapter deals with the material characterization of some of the basic types of materials used in pavements. Material characterization for soil, unbound granular material, asphalt mix, cement concrete and cemented material has been briefly discussed in this chapter.

2.2 SOIL AND UNBOUND GRANULAR MATERIAL

Compacted soil is used to build subgrade (refer Figure 2.1(a)). Unbound granular material is used to built base/sub-base of pavement structure (refer Figure 2.1(b)).

The resilient modulus (M_R) is generally the elastic modulus parameter used for characterization of granular material (or soil). It is determined by applying repetitive load to the sample in a triaxial cell [3]. The M_R is defined as

$$M_R = \frac{\text{deviatoric stress}}{\text{recoverable strain}} \tag{2.1}$$

Experimental studies indicate granular material is a stress-dependent material. A number of M_R models (as functions of state of stress) have been proposed by the past researchers. One can refer to, for example, [112,170,289,299], etc., for brief reviews on the various models of granular material and soil. One of such models is [79,120],

$$M_R = c_1\,(\sigma^c/P_a)^{c_2} \tag{2.2}$$

where σ^c is the confining pressure ($= \sigma_3$) in triaxial test, P_a is the atmospheric pressure, c_1 and c_2 are material parameters. σ^c is divided by P_a to make the parameter dimensionless. Another model, popularly known as $k-\theta$ model ($\theta = I_1$), is represented as follows [1,39,120]:

$$M_R = k_1\,(I_1)^{k_2} \tag{2.3}$$

where k_1 and k_2 are material parameters, and I_1 is the first invariant of stress (refer Equations 1.8 and 1.11). It was argued that the model represented by Equation 2.3 has certain limitations [170], and the following models were proposed [204,303]

$$M_R = k_1\,(I_1)^{k_2}\,(\sigma_d)^{k_3} \tag{2.4}$$

where k_1, k_2 and k_2 are material parameters, σ_d = deviatoric stress.

$$M_R = k_1\,(I_1)^{k_2}\,(\tau_{oct})^{k_3} \tag{2.5}$$

where τ_{oct} is the octahedral shear stress (refer Equation 1.15).

DOI: 10.1201/9781003190769-2

(a) Compacted soil used as subgrade

(b) Unbound granular material can be used as base/sub-base

Figure 2.1 Soil and unbound granular materials are used for building roads

The resilient modulus of unbound granular material depends on the number of parameters, for example aggregate gradation, level of compaction, moisture content, particle size, loading pattern, stress level, etc. Further, behaviour of granular material in compression and in tension (because granular material can only sustain small magnitude of tension), therefore these need to be modelled separately [333]. It also behaves anisotropically [261]. One can, for example, refer to [170, 289] for a brief review on this topic.

There are various other parameters which are used to characterize soil and granular materials, and some of these are related to M_R through empirical equations. Modulus of subgrade reaction [1, 225, 292] (k) is another parameter which is measured in situ typically by a plate load test [111]. This is defined as the pressure needed to cause unit displacement of the plate. This conceptually represents the 'spring constant' of the foundation on which the pavement is resting. The parameter k has been briefly discussed later in Section 3.2.1.

A large number constitutive models have been proposed for capturing the response of unbound granular materials and soils as non-linear elastic (hypo or hyper elastic), classical plasticity-based or yielding plasticity-based models. One can refer to, for example [34, 37, 166, 339], for applications of few such models in the context of pavement materials.

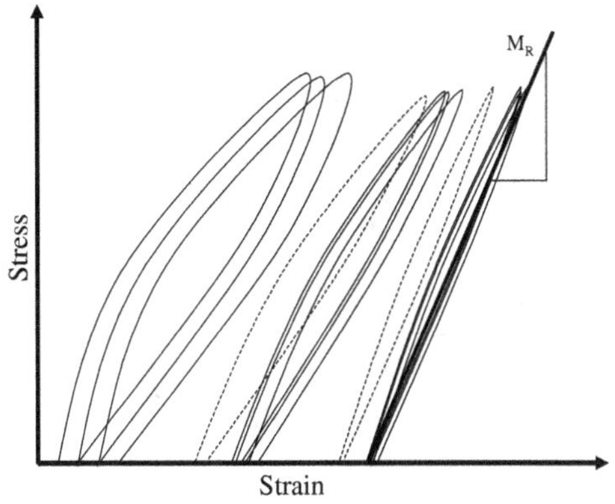

Figure 2.2 Repetitive load is applied to unbound granular material (or soil) in a triaxial set-up to estimate M_R value

2.3 ASPHALT MIX

Asphalt mix is made up of asphalt binder and aggregates mixed in specified proportion (refer Figure 2.3). Volumetrically asphalt mix contains asphalt binder, aggregates and air-voids. The suitable proportions among the constituents are decided through the process of asphalt mix design. Mix design often involves balancing between strength and volumetric criteria.

Unlike cement concrete (where hydration of cement is involved during hardening), no chemical reaction reaction takes in asphalt mix; hence, aggregates and asphalt binder retain their individual physical properties. The mixing can be done at elevated temperature (for hot mix asphalt), moderate temperature (for warm mix asphalt) or at ambient temperature (for cold mix asphalt).

Various stiffness modulus parameters, measurement techniques and models have been proposed to characterize asphalt mix [4, 30, 179, 301, 330]. Some of the simple rheological models and associated principles, which can be used to develop simple models for asphalt mix, are discussed briefly in the following.

2.3.1 RHEOLOGICAL MODELS FOR ASPHALT MIX

Rheolgical models are widely used to describe the mechanical response of materials which varies with time. For detailed understanding on rheological principles, one can refer to, for example, [60, 88, 167] etc.

Creep is a situation when stress (that is, load) is held constant, relaxation is a situation when the strain (that is, displacement) is held constant. For rheological materials like asphalt material, typically, under creep condition (that is, at constant stress situation), the strain will keep on increasing (until it stabilizes to an almost constant level), and under relaxation condition (that is, at constant strain situation), the stress will keep on decreasing (until it stabilizes to an almost constant level). A

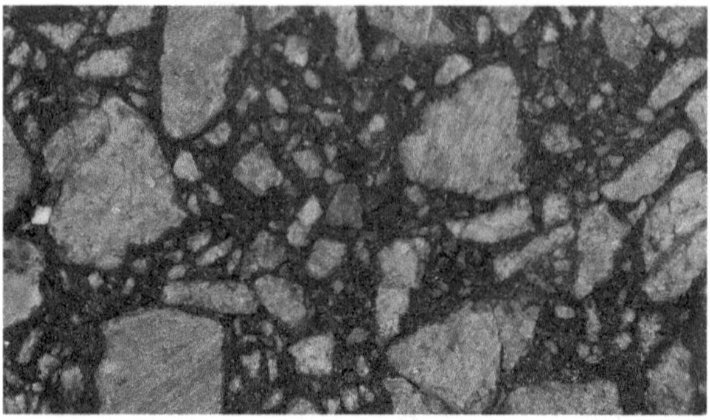

Figure 2.3 Cross-sectional image of an asphalt mix sample

typical behaviour, observed for asphalt mix, under creep and relaxation conditions is shown as Figures 2.4(a) and 2.4(b), respectively. Various curve fitting techniques (for example, power law,Prony series, etc.) have been suggested to fit the data, and the readers can, for example, refer to [154, 223] for further reading.

Creep modulus (E_{crp}) at any given time t can be defined as the stress divided by the strain at that time under creep condition, and the Creep compliance (C_{crp}) at any time t is defined as the strain divided by the stress at that time under creep

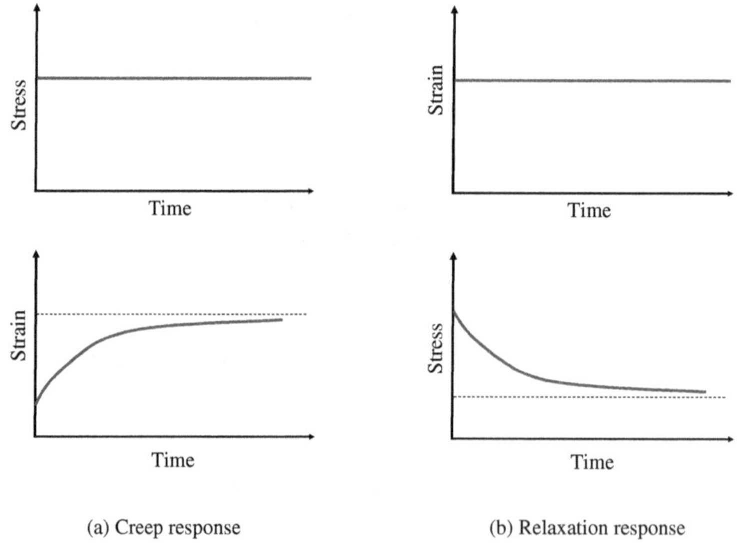

Figure 2.4 Typical creep and relaxation response of asphalt mix

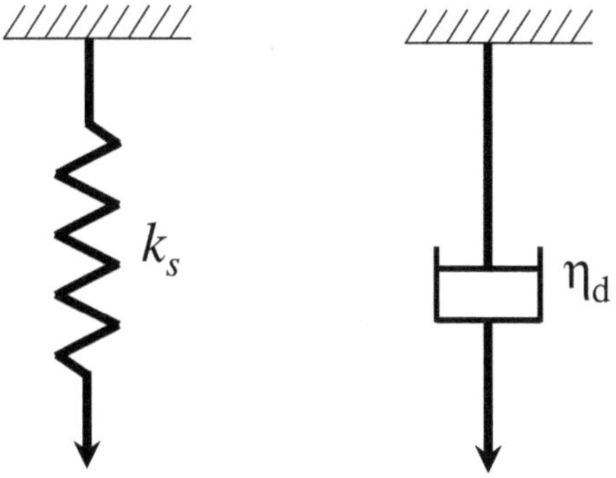

(a) A spring (b) A dashpot

Figure 2.5 A spring and a dashpot

condition. Similarly, relaxation modulus (E_{rel}) at any given time t is defined as the stress divided by the strain at that time under relaxation condition.

This behaviour can be modelled using spring(s) and dashpot(s), as axial members (as a 'mechanical analogue'), connected in various combinations. The constitutive relationship of a (Hookean) spring (refer Figure 2.5(a)) can be represented as

$$\sigma = k_s \varepsilon \tag{2.6}$$

where σ = stress in the spring, k_s = spring constant, ε = strain in the spring. The constitutive relationship of a dashpot (refer Figure 2.5(b)) can be represented as

$$\sigma = \eta_d \dot{\varepsilon} \tag{2.7}$$

where σ = stress in the dashpot, η_d = viscosity of the dashpot, $\dot{\varepsilon}$ = strain rate of the dashpot. Various combinations of spring and dashpot are used to develop various models. Since these models are made with elastic (that is, spring) and viscous (that is, dashpot) components, these are expected to capture the visco-elastic rheological behaviour of the material.

2.3.1.1 Two Component Models

A series combination of a spring and a dashpot is known as Maxwell model, and a parallel combination of a spring and a dashpot is known as Kelvin-Voigt model. These are discussed in the following.

Maxwell model

A Maxwell model is presented as Figure 2.6. Since, in this model, the spring and the dashpot are connected in series, the stresses are equal in each component and the

Figure 2.6 A Maxwell model

total strain is equal to the sum of the strains in each of the components. Considering these two conditions, the constitutive equation can be written as,

$$\sigma + \dot{\sigma}\frac{\eta_d}{k_s} = \eta_d \dot{\varepsilon} \tag{2.8}$$

where, σ = stress in the system and ε = strain in the system. For creep case, the condition is σ = constant = σ_o (say), or $\dot{\sigma} = 0$. The Equation 2.8, therefore, takes the form

$$\sigma_o = \eta_d \dot{\varepsilon} \tag{2.9}$$

Using the condition that at $t = 0$, $\varepsilon = \frac{\sigma_o}{\eta_d}$, the solution of the Equation 2.9 can be obtained as

$$\varepsilon(t) = \frac{\sigma_o}{\eta_d}t + \frac{\sigma_o}{k_s} \tag{2.10}$$

It can be seen that Equation 2.10 is a equation of a straight line and does not depict a typical creep trend (of asphalt mix) proposed in Figure 2.4(a). The creep compliance in this case can be obtained as,

$$\begin{aligned} C_{crp}(t) &= \varepsilon(t)/\sigma_o \\ &= \frac{1}{\eta_d}t + \frac{1}{k_s} \end{aligned} \tag{2.11}$$

For relaxation case, the condition is ε = constant = ε_o (say), or, $\dot{\varepsilon} = 0$. The Equation 2.8 takes the form

$$\sigma + \dot{\sigma}\frac{\eta_d}{k_s} = 0 \tag{2.12}$$

Using the condition that at $t = 0$, $\sigma = k_s\varepsilon_o$, the solution of the Equation 2.12 can be obtained as

$$\sigma = k_s \varepsilon_o e^{-\frac{k_s t}{\eta_d}} \tag{2.13}$$

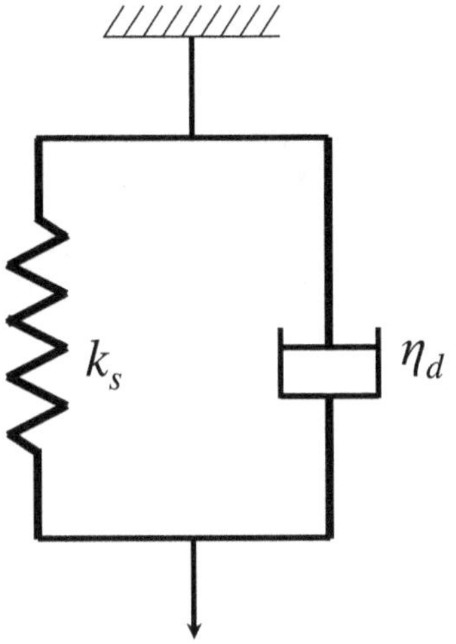

Figure 2.7 A Kelvin-Voigt model

The relaxation modulus, therefore, is obtained as

$$E_{rel}(t) = \sigma(t)/\varepsilon_o$$
$$= k_s e^{-\frac{k_s}{\eta_d}t} \tag{2.14}$$

Kelvin-Voigt model

A Kelvin-Voigt model is presented as Figure 2.7. Since, in this model, the spring and the dashpot are connected in parallel, the strains are equal in each component and the total stress is equal to the sum of the stresses in each of the components. Considering these two conditions into account, the constitutive equation can be written as

$$\sigma = k_s\varepsilon + \eta_d\dot{\varepsilon} \tag{2.15}$$

Equation 2.15 can be solved for creep condition (in a similar manner as above) and the response of the system can be obtained as the following:

$$\varepsilon = \frac{\sigma_o}{k_s}\left(1 - e^{-\frac{k_s}{\eta_d}t}\right) \tag{2.16}$$

Similarly, Equation 2.15 can be solved for relaxation condition and the response of the system can be obtained as the following:

$$\sigma = k_s\varepsilon_o \tag{2.17}$$

It can be seen that Equation 2.17 is a equation of a straight line (parallel to the time axis) and does not depict a typical relaxation trend (for asphalt mix) proposed in Figure 2.4(b).

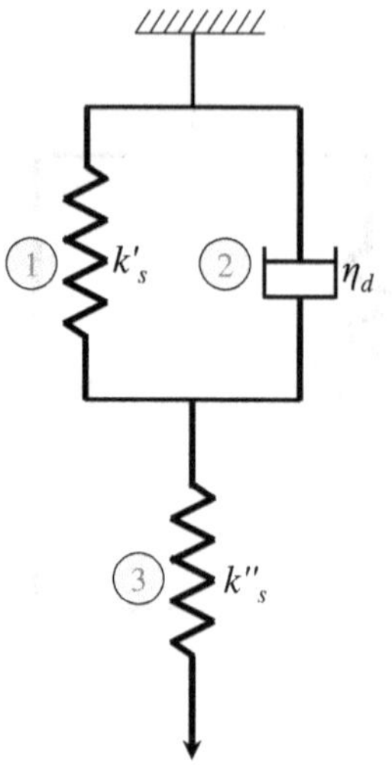

Figure 2.8 A three component model

2.3.1.2 Three Component Model

A three component model can be composed of a combination of either two springs and a dashpot or two dashpots and a spring. A model with two springs and a dashpot is called a standard solid model. A model with two dashpots and a spring is called a standard fluid model. Further depending on whether one spring and one dashpot are connected in series or in parallel, these can be called as Maxwell type or Kelvin-Voigt type.

A three component model as shown in Figure 2.8 is taken up, here, for further analysis. In this model a spring (spring constant $= k''$) is connected in series to a parallel arrangement of another spring (spring constant $= k'$) and a dashpot (viscosity $= \eta_d$). This is therefore a standard solid (Kelvin-Voigt type) model. Identifying the components as 1, 2 and 3, as shown in Figure 2.8, one can write

$$\begin{aligned} \sigma &= \sigma_3 = \sigma_1 + \sigma_2 \\ \varepsilon_1 &= \varepsilon_2 \\ \varepsilon &= \varepsilon_3 + \varepsilon_2 \\ k'_s &= \frac{\sigma_1}{\varepsilon_1} \\ \eta_d &= \frac{\sigma_2}{\dot{\varepsilon}_2} \\ k''_s &= \frac{\sigma_3}{\varepsilon_3} \end{aligned} \tag{2.18}$$

Combining all the above equations, one can obtain the constitutive relationship as

$$\sigma + \frac{\eta_d}{k'_s + k''_s}\dot{\sigma} = \frac{k'_s k''_s}{k'_s + k''_s}\varepsilon + \frac{k''_s \eta_d}{k'_s + k''_s}\dot{\varepsilon} \tag{2.19}$$

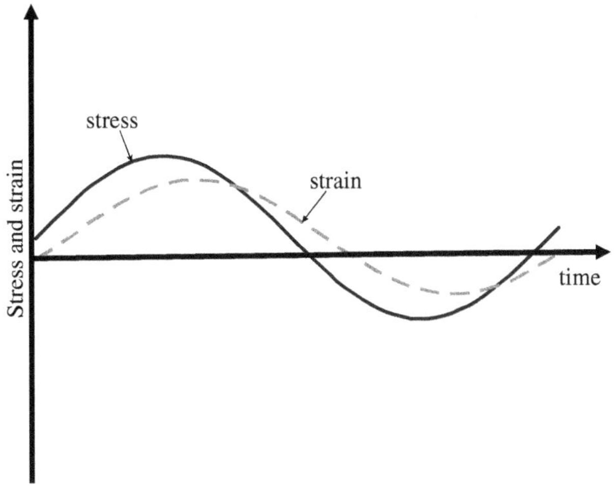

Figure 2.9 Sinusoidal loading on rheological material in stress controlled mode

The solution for the creep case (that is, $\dot{\sigma} = 0$) is obtained as

$$\varepsilon = \sigma_o \left(\frac{k_s' + k_s''}{k_s' k_s''} \left(1 - e^{-\frac{k_s'}{\eta_d} t} \right) + \frac{1}{k_s''} e^{-\frac{k_s'}{\eta_d} t} \right) \tag{2.20}$$

The solution for the relaxation case (that is, $\dot{\varepsilon} = 0$) is obtained as

$$\sigma = \varepsilon_o k_s'' e^{-\frac{k_s' + k_s''}{\eta_d} t} + \frac{k_s' k_s''}{k_s' + k_s''} \varepsilon_o \left(1 - e^{-\frac{k_s' + k_s''}{\eta_d} t} \right) \tag{2.21}$$

Under dynamic loading condition, a phase difference occurs between the stress and the strain for rheological materials. Figure 2.9 shows a schematic diagram of a stress controlled dynamic loading. The stress can be expressed as

$$\sigma = \sigma_o e^{i\omega_f t} \tag{2.22}$$

where ω_f is the angular velocity. The strain developed will have a phase angle lag of δ. That is,

$$\varepsilon = \varepsilon_o e^{i(\omega_f t - \delta)} \tag{2.23}$$

The energy dissipated in the ith cycle per unit volume (W_i) can be calculated as

$$\begin{aligned} W_i &= \int_0^{\frac{2\pi}{\omega_f}} \sigma d\varepsilon \\ &= \int_0^{\frac{2\pi}{\omega_f}} \sigma \frac{d\varepsilon}{dt} dt \\ &= i\pi \sigma_o \varepsilon_o \sin \delta \end{aligned} \tag{2.24}$$

which is the area under the hysteresis loop of the stress-strain plot and is contributed from the out-of-phase portion. For elastic material $\delta = 0$, indicating that the loss of energy due to dissipation will be zero. The complex modulus (E^*) can be written as

$$
\begin{aligned}
E^* &= \frac{\sigma}{\varepsilon} = \frac{\sigma_o e^{i\omega_f t}}{\varepsilon_o e^{i(\omega_f t - \delta)}} \\
&= \frac{\sigma_o}{\varepsilon_o} \cos\delta + i \frac{\sigma_o}{\varepsilon_o} \sin\delta \\
&= E' + iE''
\end{aligned}
\tag{2.25}
$$

where E' is defined as storage modulus, E'' as loss modulus. The dynamic modulus is defined as

$$
E_d = |E^*| = \sqrt{(E'^2 + E''^2)} \tag{2.26}
$$

The phase angle, δ, can be obtained as

$$
\delta = tan^{-1} \frac{E''}{E'} \tag{2.27}
$$

The expression for E_d can be derived for any given model. For example, incorporating $\sigma = \sigma_o e^{i\omega_f t}$ (that is, Equation 2.22) and $\varepsilon = \varepsilon_o e^{i(\omega_f t - \delta)}$ (that is, Equation 2.23) in a three-component model represented by Equation 2.19), and considering that $\dot{\sigma} = i\omega_f \sigma$ and $\dot{\varepsilon} = i\omega_f \varepsilon$, and further assuming $k'_s = k''_s = k_s$ (say), one obtains

$$
E^* = \frac{k_s^2 + i\omega_f \eta_d k_s}{2k_s + i\omega_f \eta_d} \tag{2.28}
$$

and this can be expressed in $E' + iE''$ form, as

$$
E^* = \frac{2k_s^3 + k_s \omega_f^2 \eta_d^2}{4k_s^2 + \omega_f^2 \eta_d^2} + i \frac{\omega_f \eta_d k_s^2}{4k_s^2 + \omega_f^2 \eta_d^2} \tag{2.29}
$$

Thus, the E_d (refer Equation 2.26) is obtained as

$$
E_d = |E^*| \tag{2.30}
$$

$$
= \frac{\left[(2k_s^3 + k_s \omega_f^2 \eta^2)^2 + (\omega_f \eta_d k_s^2)^2\right]^{\frac{1}{2}}}{4k_s^2 + \omega_f^2 \eta^2} \tag{2.31}
$$

and phase angle (refer Equation 2.27) is obtained as

$$
\delta = \tan^{-1}\left(\frac{\omega_f \eta_d k_s}{2k_s + \omega_f^2 \eta_d^2}\right) \tag{2.32}
$$

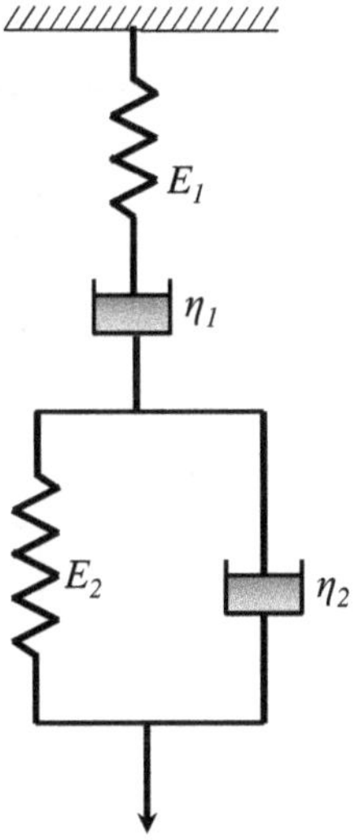

Figure 2.10 Burgers model

The E_d value of asphalt mix is dependent on a number parameters, for example, aggregate gradation, asphalt binder viscosity, temperature, volumetric parameters, level of compaction and so on. A number of predictive models have been developed to estimate the E_d value of the asphalt mix from the known parameters. Interested readers can refer to, for example, [24, 49, 154], etc. for more details.

2.3.1.3 Models with More Than Three Components

More complex models are proposed as an effort to have better match with the experimental results throughout the entire time of observation. For example, Burgers model (Figure 2.10) is a four component model consisting of a Maxwell model and Kelvin-Voigt model connected in series. The governing equation is obtained as

$$\sigma + \left(\frac{\eta_1}{E_1} + \frac{\eta_1}{E_2} + \frac{\eta_2}{E_2}\right)\dot{\sigma} + \frac{\eta_1\eta_2}{E_1E_2}\ddot{\sigma} = \eta_1\dot{\varepsilon} + \frac{\eta_1\eta_2}{E_2}\ddot{\varepsilon} \tag{2.33}$$

The creep response is obtained[1] as

$$\varepsilon(t) = \frac{\sigma_o}{E_1} + \frac{\sigma_o}{\eta_1}t + \frac{\sigma_o}{E_2}\left(1 - e^{-\frac{E_2}{\eta_2}t}\right) \tag{2.34}$$

It is interesting to observe that the creep response is the sum of creep response of Maxwell and kelvin-Voigt models.

[1]with initial conditions $\varepsilon = \frac{\sigma_o}{E_1}$ and $\dot{\varepsilon} = \frac{\sigma_o}{\eta_1} + \frac{\sigma_o}{\eta_2}$ at $t = 0$

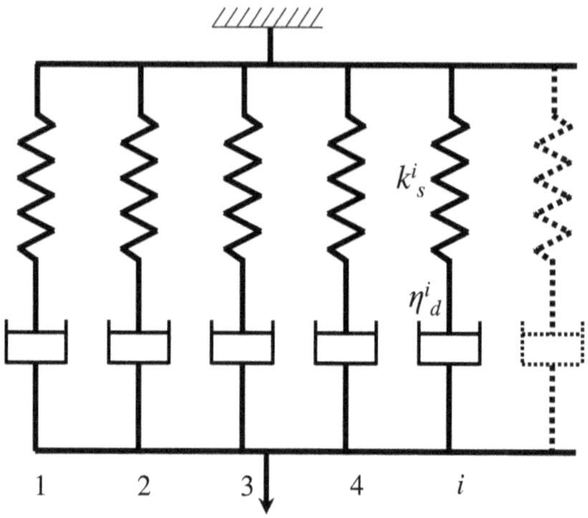

Figure 2.11 A generalized Maxwell model

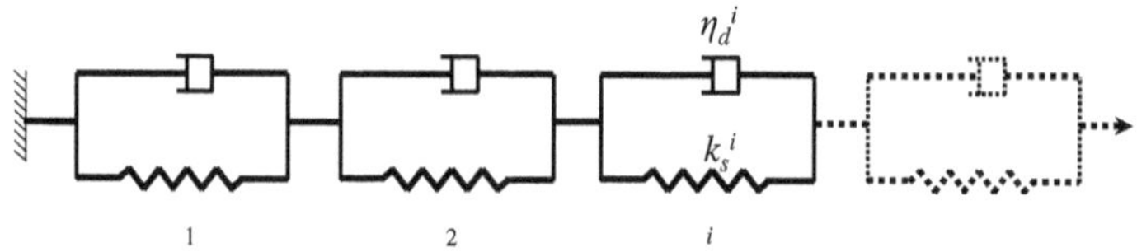

Figure 2.12 A generalized Kelvin model

In generalized models, large/infinite number of components are used. An example of generalized Maxwell model is shown in Figure 2.11. Here a number of Maxwell elements are connected in parallel. For this model, for relaxation case, one can write

$$\begin{aligned} \varepsilon_1 = \varepsilon_2 = \cdots = \varepsilon_i = \cdots = \varepsilon_o \\ \sigma = \sigma_1 + \sigma_2 + \cdots + \sigma_i + \cdots \end{aligned} \tag{2.35}$$

Thus, the relaxation response can be written as

$$\sigma(t) = \sum_{\forall i} \sigma_i = \varepsilon_o \sum_{\forall i} k_s^i e^{\frac{-k_s^i}{\eta_d^i} t} \tag{2.36}$$

One can use various other components to the model, for example, if another spring is added in parallel, it is known as Wiechert model, or Maxwell-Wiechert model.

An examples generalized Kelvin model is shown in Figure 2.12[2]. For this model, for creep case, one can write

$$\begin{aligned} \sigma_1 &= \sigma_2 = \cdots = \sigma_i = \cdots = \sigma_o \\ \varepsilon &= \varepsilon_1 + \varepsilon_2 + \cdots + \varepsilon_i + \cdots \end{aligned} \tag{2.37}$$

[2]Variants are possible and one can add various other components.

Thus, the creep response can be written as

$$\varepsilon(t) = \sum_{\forall i} \varepsilon_i = \sigma_o \sum_{\forall i} \frac{1}{k_s^i} \left(1 - e^{-\frac{k_s^i}{\eta_d^i} t}\right) \quad (2.38)$$

Other than the above examples of two-component, three-component and generalized models discussed, various other combinations of components are possible and accordingly various models have been proposed (for example, Huet model, Huet-Sayegh model and so on). Researchers use various such models to capture the time-dependent behaviour of asphaltic material. One can refer to, for example [64, 154] for further study.

2.3.1.4 Linear Viscoelasticity

A rheologoical material is linearly visco-elastic when it follows the conditions of homogeneity and linear superposition. These two conditions are explained in the following:

- Homogeneity: For a linearly viscoelastic material, if the stress applied (that is, for a stress controlled experiment) is increased by a factor (than what was applied in the first experiment), the strain response at any given time will also increase by the same factor (than the first experiment at the same time). This is the homogeneity condition. This is schematically explained in Figure 2.13. Similarly, if the strain applied (for a strain controlled experiment) is increased by a factor, the stress response at any given time will also increase by the same factor.
- Superposition: For a linearly viscoelastic material, the overall response (due to different loads applied at different points of time) can be obtained by linear superposition of the individual responses (due to the same load individually applied at respective points of times). This is the superposition condition. This is schematically explained in Figure 2.14 for a hypothetical stress controlled experiment. In Figure 2.14(a) a stress of σ_1 is applied at time $t = \xi_1$, and its strain response is shown in the diagram below. In Figure 2.14(b) another stress of σ_2 is applied[1] at time $t = \xi_2$, and its strain response is shown in the diagram below. The principle of superposition suggests that the strain response will be the same as the sum of the strain response of the above two cases, if stress, both σ_1 and σ_2 are applied on the material at times ξ_1 and ξ_2, respectively (refer to Figure 2.14(c)).

The above conditions are utilized to derive Boltzmann's linear superposition principle. Figure 2.15 shows a strain controlled thought-experiment on any rheological material, like asphalt mix. In this experiment, an incremental strain of $\Delta\varepsilon_1$ is applied at time $t = \zeta_1$, then another incremental of strain $\Delta\varepsilon_2$ is applied at time $t = \zeta_2$ and so on.

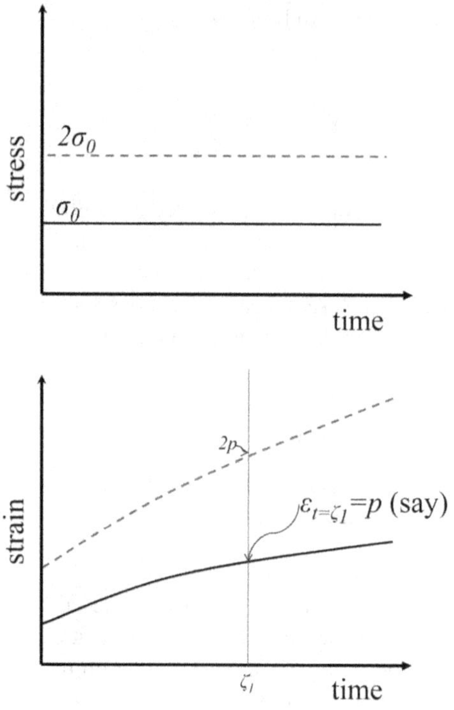

Figure 2.13 A schematic diagram explaining the principle of homogeneity

If these cause increments of stress by $\Delta\sigma_1$, at time $t = \zeta_1$ and then, $\Delta\sigma_2$ at $t = \zeta_2$ and so on, then, by using the above-mentioned conditions of linear viscoelasticity, one can add for all these time steps to obtain

$$\begin{aligned} \Delta\sigma_1 &= E_{rel}\,(t-\zeta_1)\,\Delta\varepsilon_1 \\ \Delta\sigma_2 &= E_{rel}\,(t-\zeta_2)\,\Delta\varepsilon_2 \end{aligned}$$

and so on.

Assuming that the initial stress level of the material as zero (that is, $\sigma|_{t=0} = 0$) in this experiment, one can write

$$\sigma(t) = \sigma|_{t=0} + \sum_{\forall i} E_{rel}(t-\zeta_i)\Delta\varepsilon_i \tag{2.39}$$

Thus, if a varying strain is applied to a rheological material, it can be discretized into small time steps, and the above formulation (Equation 2.39) can be used to obtain the stress at a time t. In a similar manner, for stress controlled experiment, one can derive

$$\varepsilon(t) = \varepsilon|_{t=0} + \sum_{\forall i} C_{crp}(t-\zeta_i)\Delta\sigma_i \tag{2.40}$$

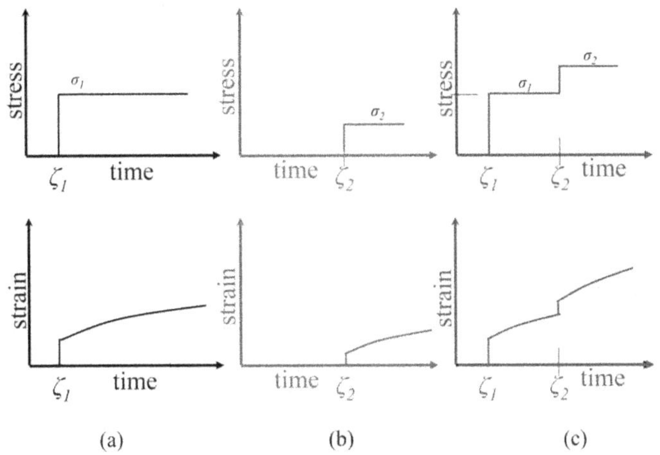

Figure 2.14 A schematic diagram explaining the principle of linear superposition

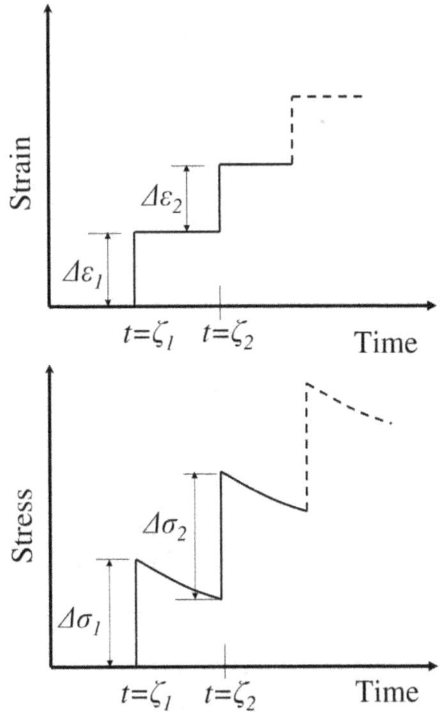

Figure 2.15 A strain controlled thought experiment

For a continuous domain, these equations can be equivalently written as

$$\sigma(t) = \int_0^t E_{rel}(t-\zeta)\frac{d\varepsilon(\zeta)}{d\zeta}d\zeta \quad \text{(for strain controlled case)} \tag{2.41}$$

$$\varepsilon(t) = \int_0^t C_{crp}(t-\zeta)\frac{d\sigma(\zeta)}{d\zeta}d\zeta \quad \text{(for stress controlled case)} \tag{2.42}$$

This is known as Boltzmann's superposition principle for linear visco-elastic material. By taking Laplace transform to Equations 2.41 and (and using the product and derivative theorems of Laplace transform and assuming $\varepsilon|_{t=0} = 0$) one obtains [88]

$$\overline{\sigma}(s) = s\overline{E_{rel}}(s)\overline{\varepsilon}(s) \tag{2.43}$$

Similarly, from Equation 2.42 one obtains [88]

$$\overline{\varepsilon}(s) = s\overline{C_{crp}}(s)\overline{\sigma}(s) \tag{2.44}$$

where $\overline{\sigma}(s)$, $\overline{\varepsilon}(s)$, $\overline{E_{rel}}(s)$, $\overline{C_{crp}}(s)$ are the stress, strain, relaxation modulus and creep modulus in the Laplacian domain.

Equations 2.43 and 2.44 look analogous to equations for elastic material (with the modulus term multiplied with s), but in Lapacian domain. This principle is thus called as elastic analogy of viscoelastic material, and also known as elastic-viscoelastic correspondence principle.

Further, from Equations 2.43 and 2.44, one can write

$$\overline{C_{crp}}(s)\,\overline{E_{rel}}(s) = \frac{1}{s^2} \tag{2.45}$$

Further, it is interesting to note that in general $E_{rel} \neq \frac{1}{C_{crp}}$ (except for some special situations), but, C_{crp} and E_{rel}, for linearly viscoelastic material maintains the above (given by, Equation 2.45) relationship in Laplacian domain.

Equation 2.45 provides a link between the moduli value obtained from creep and relaxation tests for linear viscoelastic material. For a given expression for creep response (say), it may be possible to find out the expression for relaxation response, without using the component structure of the model. Some illustrative problems are presented below. One may refer to, for example, [88, 167, 213] for further studies on the concepts of linear visco-elasticity and various inter-relationships that may exist amongst the parameters.

Example problem

A stress of magnitude σ_o is applied at $t = \zeta_1$ on a rheological material (which was initially stress-free) given as Equation 2.15, then withdrawn at $t = \zeta_2$ (refer Figure 2.16). Predict the strain response.

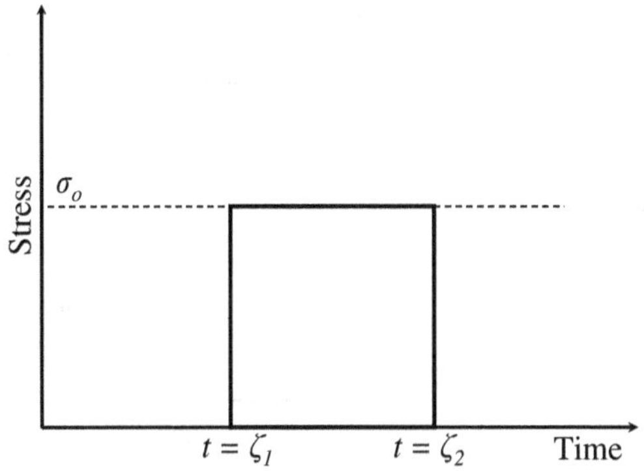

Figure 2.16 A stress of magnitude σ_o applied at $t = \zeta_1$ on a rheological material, then withdrawn at $t = \zeta_2$

Solution

Equation 2.15 represents a Kelvin-Voigt model, whose creep response is given by Equation 2.16. From Equation 2.16, one can write

$$C_{crp} = \frac{1}{k_s}\left(1 - e^{-\frac{k_s t}{\eta_d}}\right) \tag{2.46}$$

Using Equation 2.40, one can write

$$\varepsilon(t) = \frac{\sigma_o}{k_s}\left(1 - e^{-\frac{k_s(t-\zeta_1)}{\eta_d}}\right) \quad \text{for } \zeta_2 < t < \zeta_1 \tag{2.47}$$

$$\varepsilon(t) = \frac{\sigma_o}{k_s}\left(e^{-\frac{k_s(t-\zeta_2)}{\eta_d}} - e^{-\frac{k_s(t-\zeta_1)}{\eta_d}}\right) \quad \text{for } t > \zeta_2 \tag{2.48}$$

The above strain response is schematically shown in Figure 2.17. Since Kelvin-Voigt model has a spring and a dashpot connected in parallel (and no spring in series), the strain response does not show any (i) instantaneous deformation on application of stress and (ii) instantaneous recovery on withdrawal of stress (refer Figure 2.17).

Example problem

A stress of magnitude σ_o is applied (which was initially stress-free) over a rheological material over a period of ζ_1 as shown in Figure 2.18. If $C_{crp} = \left(\frac{t}{\eta_d} + \frac{1}{k_s}\right)$ for the material, estimate the strain at time t, for $t > \zeta_1$.

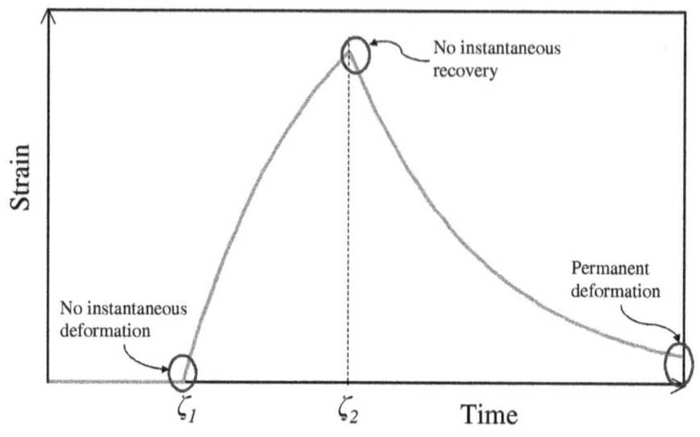

Figure 2.17 Strain response of Kelvin-Voigt model when a stress of magnitude σ_o applied at $t = \zeta_1$ and then withdrawn at $t = \zeta_2$

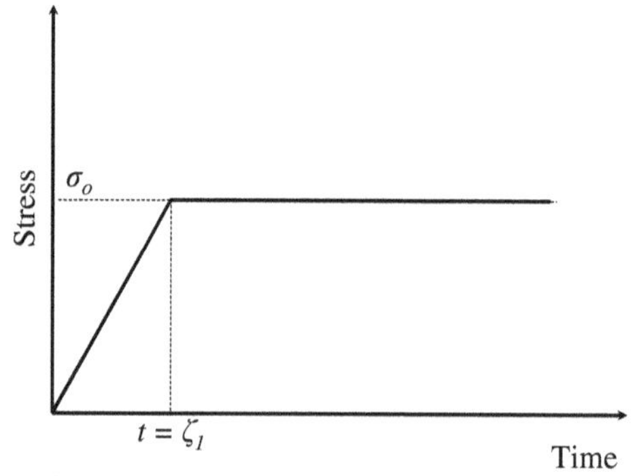

Figure 2.18 A stress of magnitude σ_o is applied linearly by $t = \zeta_1$ on a rheological material

Solution

The strain at time t (where, $t > \zeta_1$) is given as

$$\begin{aligned}\varepsilon_t|_{t>\zeta_1} &= \int_0^{\zeta_1}\left(\frac{t-\zeta}{\eta_d}+\frac{1}{k_s}\right)\frac{\sigma_o}{\zeta_1}d\zeta + \int_{\zeta_1}^{t}\left(\frac{t-\zeta}{\eta_d}+\frac{1}{k_s}\right)0\,d\zeta \\ &= \sigma_o\left(\frac{(t-\zeta_1/2)}{\eta_d}+\frac{1}{k_d}\right)\end{aligned} \tag{2.49}$$

Example problem

Creep compliance of a rheological model is given as

$$C_{crp}(t) = \frac{t}{\eta_d} + \frac{1}{k_s}$$

Find out an expression for the relaxation modulus ($E_{rel}(t)$).

Solution

After taking Laplace transformation, one obtains

$$\overline{C_{crp}}(s) = \frac{1}{\eta_d s^2} + \frac{1}{k_s s}$$

Using Equation 2.45, one obtains

$$\begin{aligned}\overline{E_{rel}}(s) &= \frac{1}{s^2 C_{crp}(s)} \\ &= \frac{k_s}{\frac{k_s}{\eta} + s}\end{aligned}$$

By taking inverse Laplace transform, one obtains

$$E_{rel}(t) = k_s e^{\frac{-k_s}{\eta} t}$$

Example problem

A load of $\sigma(t) = \sigma_o e^{i\omega_f t}$ is applied on a three component model shown in Figure 2.8). The creep response of the model is given as Equation 2.20. Calculate the E^* using Boltzmann's superposition principle. Assume $k_s' = k_s'' = k_s$.

Solution

Considering Equation 2.20 and assuming $k_s' = k_s'' = k_s$

$$\begin{aligned}C_{crp} &= \left(\frac{2}{k_s}\left(1 - e^{-\frac{k_s}{\eta_d} t}\right) + \frac{1}{k_s} e^{-\frac{k_s}{\eta_d} t}\right) \\ &= \frac{1}{k_s}\left(2 - e^{-\frac{k_s}{\eta_d} t}\right)\end{aligned} \tag{2.50}$$

From Equation 2.42, one can write

$$\begin{aligned}\varepsilon(t) &= \int_0^t \frac{1}{k_s}\left(2 - e^{-\frac{k_s}{\eta_d}(t-\zeta)}\right)(i\omega_f)\sigma_o e^{i\omega_f \zeta} d\zeta \\ &= \frac{\sigma_o e^{i\omega_f t}}{k_s}\left(2 - \frac{i\omega_f}{\frac{k_s}{\eta_d} + i\omega_f}\right) \\ &= \frac{\sigma(t)}{k_s} \frac{(2k_s + i\omega_f \eta_d)}{(k_s + i\omega_f \eta_d)}\end{aligned} \tag{2.51}$$

Or,

$$\frac{\sigma(t)}{\varepsilon(t)} = \frac{k_s^2 + i\omega_f \eta_d k_s}{2k_s + i\omega_f \eta_d} = E^* \tag{2.52}$$

It may be noted that the above expression for E^* (that is, Equation 2.52) obtained using Boltzmann's superposition principle is same as Equation 2.28.

2.3.1.5 Time-Temperature Superposition

The response of asphaltic material shows dependency on time as well on temperature. For example, variation of creep compliance at different time and at two temperatures (say, T' and T'', where, $T' > T''$) has been plotted schematically in Figure 2.19. From Figure 2.19, it can be seen that at a given temperature, say at T'', the C_{crp} at time t' is lower than that of time t'', (i.e. $C'_{crp} < C''_{crp}$) – this is because, as time increases the strain keeps on increasing. Further, it can be seen that at a given time, say at t', the C_{crp} at temperature T'' is lower than that of temperature T' (i.e. $C'_{crp} < C''_{crp}$) – this is because, if the temperature is higher the strain will be more at the same given time.

This behaviour speaks for an equivalency that may exist between time and temperature, for example, the C_{crp} measured at time t'' at temperature T'' is equal to the C_{crp} measured at time t' at temperature T' (and its value is C''_{crp} as shown in Figure 2.19). That means, one may perform a test at specific temperature and time to obtain a rheological parameter for some other temperature and time. This forms the basis of time-temperature superposition.

Thus, a relationship between the two time scales can be proposed as

$$t' = \frac{t''}{\alpha_T} \tag{2.53}$$

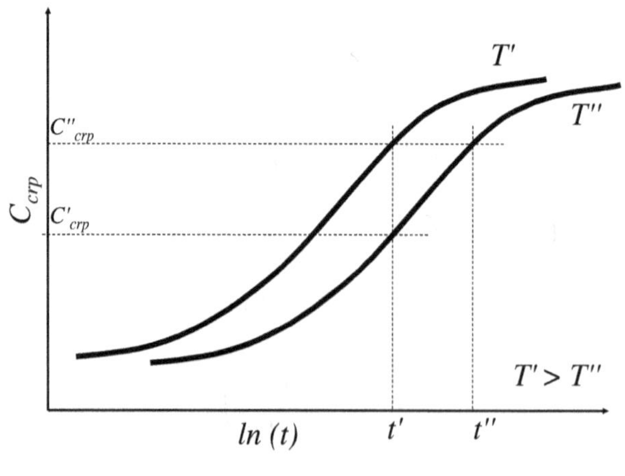

Figure 2.19 Schematic diagram illustrating the principle of time-temperature superposition

$$\text{Or, } ln(t') = ln(t'') - ln(\alpha_T) \tag{2.54}$$

where t' can be called as reduced time of t'' for shifting the test temperature from T'' to T', and α_T is called the time-temperature shift factor. Obviously, α_T is a function of these two temperatures, one of them can be treated as a standard temperature. Materials for which α_T value is not dependent on time are called as thermorheologically simple material.

Certain formulations are proposed for calculation of α_T for thermo-rheologically simple material. The following two formulae are typically used for asphalt mixes. The Williams-Landel-Ferry (WLF) equation is given as [322],

$$log_e(\alpha_T) = \frac{-C_1\,(T - T_{\text{ref}})}{C_2 + (T - T_{\text{ref}})} \tag{2.55}$$

where $log_e(\alpha_T)$ is the natural logarithm of α_T, T_{ref} is the reference temperature, T is a temperature where α_T is being determined, C_1 and C_2 are constants.

The Arrhenius equation [90, 331] is given as

$$log_e(\alpha_T) = \frac{\Delta H}{log_e(10)U_g}\left(\frac{1}{T} - \frac{1}{T_{\text{ref}}}\right) \tag{2.56}$$

where ΔH is the apparent activation energy, and U_g is the universal gas constant.

If rheological tests are conducted at various temperatures[3], it is possible (and, more easily for a thermo-rheologically simple materials) to develop the complete spectrum of rheologoical behaviour of the material at any specified reference temperature (T_{ref}). This is known as master curve. Further, for dynamic testing it is possible to establish equivalency with the frequency of loading (w_f) to time (for tests with static loading conditions) or temperature. Figure 2.20 schematically shows a dynamic modulus versus frequency master curve. The data obtained for the tests conducted at temperatures T', T'' and T''' are shifted to a reference temperature T_{ref} to develop this master curve. One can, for example, refer to [154, 213, 227, 256, 301] on the (i) development of master curve, (ii) interconversion between time, temperature, frequency and (iii) techniques to obtain the curve fit parameters.

Example problem

Creep test was performed on a thermo-rheologically simple visco-elastic material after the sample was kept in two different conditions one after the other. First, the sample was kept at 10^oC for 10 hours, next it was switched to 100^oC and was kept for 10 minutes and immediately after this the creep modulus value is noted.

Find the equivalent time (reduced time) at which the test should be performed if the sample is decided to be kept at a constant temperature of (i) 10^oC or (i) 100^oC for the entire period, so as to obtain the same value of creep modulus.

Given $log_{10}\alpha_T = \pm 2$

[3] depending on the test convenience and available experimental facilities.

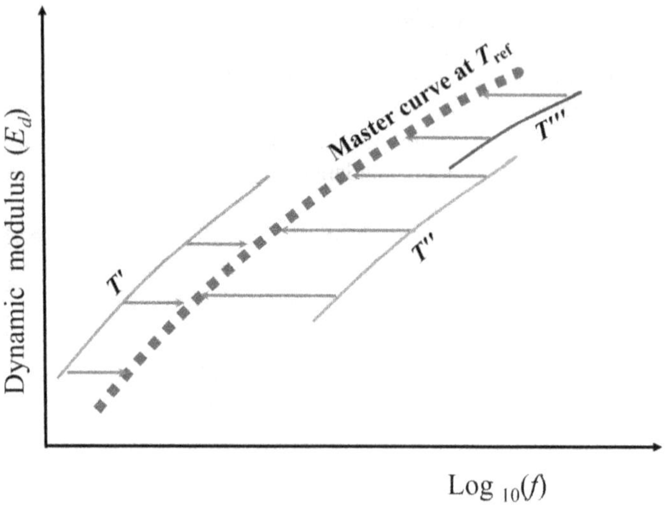

Figure 2.20 Development of master curve

Solution

(i) If the temperature is maintained as 10°C throughout, then creep modulus should be measured at 10 hours $+\frac{10}{10^{-2}}$ minutes $= 26.67$ hours.

(ii) If the temperature is maintained as 100°C throughout, then creep modulus should be measured at $\frac{10}{10^{+2}}$ hours $+$ 10 minutes $=$ 16 minutes.

2.3.1.6 Discussions

Simple rheological models have been presented in the above, and some of these are used as descriptor of rheological behaviour of asphalt mix [332, 340]. However, asphalt mix, as well as asphalt binder, shows much complex behaviour in terms of their dependency on time, temperature and stress state. Further, asphalt mix is of anisotropic [302] and heterogeneous [14] materials. Researchers have been trying to develop various models to capture the response and damage mechanism of this complex material [22, 155, 163, 250]. Interested readers may refer to [154, 162, 335] for a review and further study on mechanical modelling of asphalt mix (and asphalt binder).

2.3.2 VISCOELASTIC POISSON'S RATIO

The ratio of axial strain to the transverse strain is the Poisson's ratio. However, for viscoelastic material, these strains are time-dependent, and also dependent on the mode of loading [7, 145].

For elastic isotropic solid, one can write

$$\nu = \left[\frac{1}{2} - \frac{E}{6K}\right] \tag{2.57}$$

where ν is the Poisson's ratio, E is the Young's modulus, K is the bulk modulus.

Given that, $\nu = -\frac{\varepsilon_{tr}}{\varepsilon_L}$, where ε_{tr} is the transverse strain and ε_L is the longitudinal strain.

For an elastic uniaxial member, one can write

$$\varepsilon_{tr} = -\varepsilon_L \left[\frac{1}{2} - \frac{E}{6K}\right] \tag{2.58}$$

Consider a strain-controlled uniaxial loading on a linearly viscoelastic material. Using elastic-viscoelastic correspondence principle (refer to Section 2.3.1.4), one can write [168]

$$\overline{\varepsilon_{tr}}(s) = -\left[\frac{1}{2}\overline{\varepsilon_L}(s) - \frac{s\overline{E_{rel}}(s)\overline{\varepsilon_L}(s)}{6K}\right] \tag{2.59}$$

This is assuming that the bulk modulus remains is constant over time. By taking inverse Laplace transform of Equation 2.59, one obtains [168]

$$\begin{aligned}
\varepsilon_{tr}(t) &= -\left[\frac{1}{2}\varepsilon_L(t) - \frac{1}{6K}\int_0^t E_{rel}(t-\zeta)\frac{d\varepsilon(\zeta)}{d\zeta}d\zeta\right] \\
&= -\left[\frac{1}{2}\varepsilon_L(t) - \frac{1}{6K}\sigma_L(t)\right] \\
&= -\varepsilon_L(t)\left[\frac{1}{2} - \frac{1}{6K}E_{rel}(t)\right]
\end{aligned} \tag{2.60}$$

Thus,

$$\nu(t) = -\frac{\varepsilon_{tr}(t)}{\varepsilon_L(t)} = \frac{1}{2} - \frac{1}{6K}E_{rel}(t) \tag{2.61}$$

One can refer to [168] for the derivation for the case K is does not remain constant. The above expression utilizes the bulk modulus (K), under a plane stress situation (2-D), one may proceed in the following way [153]. Considering elastic-viscoelastic correspondence for the expressions of ε_{xx} and ε_{zz} in Equation1.32, one can write

$$\begin{aligned}
\overline{\varepsilon_{xx}}(s) &= s\overline{C_{crp}}(s)\left(\overline{\sigma_{xx}}(s) - s\overline{\nu}(s)\overline{\sigma_{zz}}(s)\right) \\
\overline{\varepsilon_{zz}}(s) &= s\overline{C_{crp}}(s)\left(\overline{\sigma_{zz}}(s) - s\overline{\nu}(s)\overline{\sigma_{xx}}(s)\right)
\end{aligned} \tag{2.62}$$

From Equation 2.62, one gets [153]

$$\nu(t) = L^{-1}\left[\frac{1}{s}\left(\frac{\overline{\varepsilon_{zz}}(s)\overline{\sigma_{xx}}(s) - \overline{\varepsilon_{xx}}(s)\overline{\sigma_{zz}}(s)}{\overline{\varepsilon_{zz}}(s)\overline{\sigma_{zz}}(s) - \overline{\varepsilon_{xx}}(s)\overline{\sigma_{xx}}(s)}\right)\right] \tag{2.63}$$

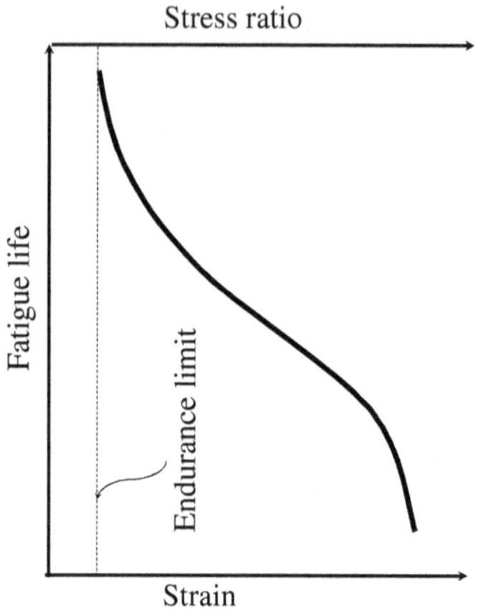

Figure 2.21 Schematic diagram showing a possible fatigue behaviour of bound materials in simple flexural fatigue testing

Further, derivations can be developed for the expressions of viscoelastic Poisson's ratio for uniaxial creep, dynamic loading, bending, etc., and their interrelationships [168].

Considerable literature is available on the choice of Poisson's ratio value of pavement materials [1, 39, 145, 210, 215]. For pavement design purpose, for a given pavement material, generally a constant value of Poisson's ratio is assumed. It is generally considered that variation of Poisson's ratio values does not significantly affect the pavement analysis results [210].

2.3.3 FATIGUE CHARACTERIZATION

Bound materials (like, asphalt mix) undergo fatigue damage due to repetitive application of load. In the laboratory, the loading may be applied in a stress or strain controlled manner on samples of various geometries. The repetitive loading may be flexural, axial or torsional in nature; however, flexural fatigue loading is generally used for pavement engineering applications. Loading can be simple (where stress or strain amplitude level is maintained constant) or compound (where stress or strain amplitude level is varied during the course of testing) in nature.

Figure 2.21 shows a typical fatigue characteristics of bound materials due to simple flexural fatigue loading. The fatigue life is defined as the number of load repetitions (cycles) at the which the material deteriorates to some pre-defined level. The researchers have proposed various definitions (and approaches) for estimation of fatigue life – a brief overview of these are presented towards the later part of this section. The stress ratio is defined as the ratio between the applied stress amplitude (for constant stress amplitude testing) and the flexural strength (known as modulus of rupture) of the bound material. Strain or stress ratio is generally used for constant strain amplitude or constant stress amplitude fatigue testing, respectively.

Typically, strain controlled test is performed for fatigue characterization of asphalt material because of the ductile nature of the material. And stress controlled test is performed for fatigue characterization of cement concrete or cemented material, because of the brittle nature of the material.

From Figure 2.21, it can be seen that if the strain amplitude (or stress ratio) level is high, the fatigue life is expected to be low and vice versa. It has been observed that at a very low level of strain (or stress ratio), the sample does not fail due to such repetitive loading[4] and this is known as endurance limit. This property later formed the basis of perpetual pavement design [217,314].

For compound fatigue loading, the following empirical relationship generally satisfies

$$\sum_{\forall i} \frac{n_i}{N_i} \approx 1 \tag{2.64}$$

where n_i = the number of load repetitions (cycles) applied at a given strain (or stress ratio) level, and N_i is the fatigue life of the material at that strain (or stress ratio) level. Equation 2.64 was originally developed based on experiments conducted on aluminium [209], and subsequently adopted in pavement engineering for characterizing fatigue behaviour of asphalt mix, cement concrete and cemented material [74,106,130,214,225,281,292]. By using Equation 2.64, one assumes that fractional damages caused due to repetitions at various levels of strains (or stresses) are linearly accumulative [86].

As mentioned, various alternative definitions have been proposed for fatigue life [26]. These can be summarized as follows:

- In one approach the fatigue life is defined as the number of cycles at which the elastic modulus (say, dynamic modulus) reaches a predefined fraction (say 50%) of the original elastic modulus value [2, 183, 281]. Figure 2.22 shows possible variation of stiffness with the loading cycles. Depending on the rate of decrease of stiffness, the diagram can be divided into a number of zones as shown in the figure [5]. The basic equation, for constant strain amplitude fatigue testing, proposed is

$$N_{lf} = k_1 \left(\frac{1}{\varepsilon}\right)^{k_2} \tag{2.65}$$

  where N_{lf} is the fatigue life, ε is the (bending) strain amplitude, k_1 and k_2 are constants obtained through regression.

[4]that is, the fatigue life in-principle becomes infinity.

[5]However, depending on the test conditions, sample geometry and type of loading these zones may vary.

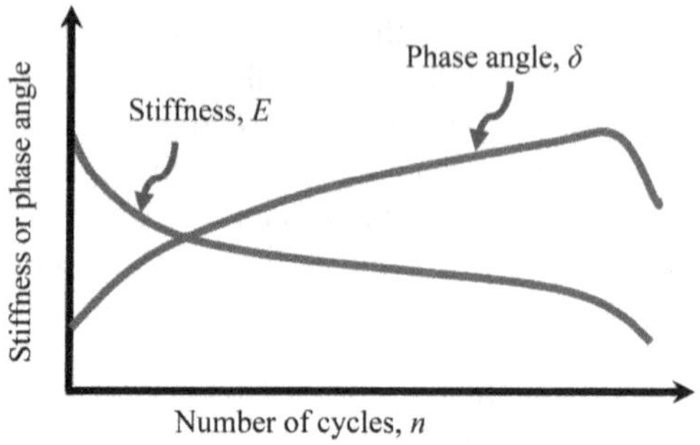

Figure 2.22 Schematic diagram showing diagram possible variation of stiffness and phase angle with loading cycles during asphalt beam fatigue testing

- In another approach once a specific value of the phase angle is reached, the sample is said to have failed [242, 308]. Figure 2.22 also shows possible variation of phase angle with the loading cycles.
- In another approach the cumulative dissipated energy [305] is used for estimating fatigue life. The dissipated energy per cycle (of a linearly viscoelastic material) can be calculated using Equation 2.24. It is proposed that the fatigue life (N_{lf}) and the cumulative dissipated energy (W_T) can be related. A simple empirical form can be suggested as [305–307].

$$
\begin{aligned}
W_T &= \sum_{i=1}^{N_{lf}} W_i \\
&= k_3 \left(N_{lf}\right)^{k_4}
\end{aligned}
\tag{2.66}
$$

where k_3 and k_4 are constants obtained through regression. As an attempt to make this empirical formulation independent of type of loading (constant stress or strain amplitude), various adjustment factors are suggested [306]. However, this approach does not separate between the dissipated energy for damage and viscoelastic dissipation [26, 308].
- In another approach dissipated energy ratio (DER) is considered to identify the fatigue failure of the sample, and hence the fatigue life [232]. The DER is defined as

$$DER = \frac{W_T}{W_i} \tag{2.67}$$

In another approach significant change in the dissipated energy between the consecutive cycles is proposed as an indicator of failure [104], and hence the fatigue life.
- In some approaches dissipated pseudo-stiffness [251] or pseudo-strain energy [156] has been used as parameters to identify fatigue life. Pseudo strain

is defined as

$$\varepsilon_R = \frac{1}{E_R}\int_0^t E_{rel}(t-\zeta)\frac{d\varepsilon(\zeta)}{d\zeta}d\zeta \tag{2.68}$$

in a strain-controlled loading situation, where E_R is the reference elastic modulus. The pseudo-stiffness, $C(t)$ is given as

$$C(t) = \frac{\sigma(t)}{\varepsilon_R} \tag{2.69}$$

Significantly large number of research studies are available on fatigue characterization of asphaltic material, covering the issues related to mode/process of testing [2, 74, 201, 240, 281], factors affecting the fatigue behaviour [30, 183, 240], variability in test results [74, 240, 281], fracture and damage modelling of asphalt mix [66, 102, 155], stiffness reduction [2, 183], asphalt healing [20, 157], endurance limit [323], and so on. One can refer to articles, like [19, 162, 184] for an overview on fatigue behaviour of asphalt mix.

2.4 CEMENT CONCRETE AND CEMENTED MATERIAL

Cement concrete is made up of aggregates, cement, admixtures (if required) and water. The suitable proportions among the constituents are decided through the process of cement concrete mix design. Mix design often involves balancing between various strength and workability criteria. Hydration of cement and subsequent hardening contributes to the strength of the material. Cement concrete, in hardened state, is characterized by its elastic modulus, compressive strength, tensile strength, bending strength (that is, modulus of rupture), etc. Empirical equations are suggested for interrelationships between these physical properties [5, 215, 216]. Elastic modulus of cement concrete can be measured as tangent modulus, secant modulus or dynamic modulus [215]. Fatigue performance of cement concrete is an important consideration in the concrete pavement design [1, 136, 214, 220, 225, 292, 296]. One can, for example, refer to text books such as [215, 216], etc., for a detailed study on the properties and characterization of cement concrete.

Cemented materials (generally locally available/marginal materials are utilized as bound form) are bound material; hence, these can be characterized in a similar manner, like other bound material. Interested readers can refer to, for example [91, 116, 176, 225], for further study on the characterization of cemented material and their application in pavement construction.

2.5 CLOSURE

This chapter has discussed the concepts of 'elastic modulus' (stiffness) of various materials used for road construction. The elastic moduli of materials are used as an essential input during pavement analyses, which are discussed in subsequent chapters (Chapters 3 to 6). The issues related to time dependency (for asphaltic material) and stress dependency (for unbound granular material) of elastic modulus have been highlighted. Pavement layers undergo damage (for example, damage due to fatigue for bound materials, permanent deformation, etc.). Generally, such damages propagate with the load repetitions (cycles). Number of repetitions a pavement can sustain until failure is an important consideration in pavement design. This has been discussed further in Chapter 8.

3 Load Stress in Concrete Pavement

3.1 INTRODUCTION

Concrete pavements are often idealized as beams (1-D) or plates (2-D) on elastic foundation (refer Figure 3.1). Numerous hypothetical springs placed at the bottom of this beam/plate represent the elastic foundation for such models (not shown in Figure 3.1). This foundation represent the combined support provided by base/sub-base and subgrade. In this chapter first the analysis of a beam resting on an elastic foundation is presented followed by the analysis of a thin plate resting on elastic foundation. Further, it is discussed how the plate theory can be utilized for the analysis of an isolated concrete pavement slab (of finite dimension) resting on a base/sub-base.

3.2 ANALYSIS OF BEAM RESTING ON ELASTIC FOUNDATION

Beam is a one-dimensional (1-D) member. Schematic diagram of a beam (say, of unit width) resting on numerous closely spaced springs with spring constant k (known as Winkler's model, discussed later in Section 3.2.1) subjected to a pointed loading Q at $x = 0$ is presented as Figure 3.2. The free-body diagram of a portion of the beam (other than the load application point) is shown in Figure 3.3. The moment (M) and shear force (V) are shown on the free-body diagram[1]. The upward force of the spring on the element of length dx is $wkdx$, where w is the displacement of the beam in the Z direction. From Figure 3.3, using force equilibrium, one can write

$$V - (V + dV) + kwdx = 0 \tag{3.1}$$

Taking moment equilibrium, one obtains

$$V = \frac{dM}{dx} \tag{3.2}$$

Putting Equation 3.2 in Equation 3.1 and considering that $EI\frac{d^2w}{dx^2} = -M$, where E is the elastic modulus of the beam and I is the second moment or area, thus EI is the flexural rigidity of the beam (that is, for Euler–Bernoulli beam and for the coordinate system chosen) one obtains

$$EI\frac{d^4w}{dx^4} + kw = 0 \tag{3.3}$$

This is a widely used basic expression for beam resting on elastic foundation applied to foundation engineering [119, 148, 258]. The general solution of Equation 3.3 is

[1]The directions of moment and shear on the positive face of the element are considered positive.

DOI: 10.1201/9781003190769-3

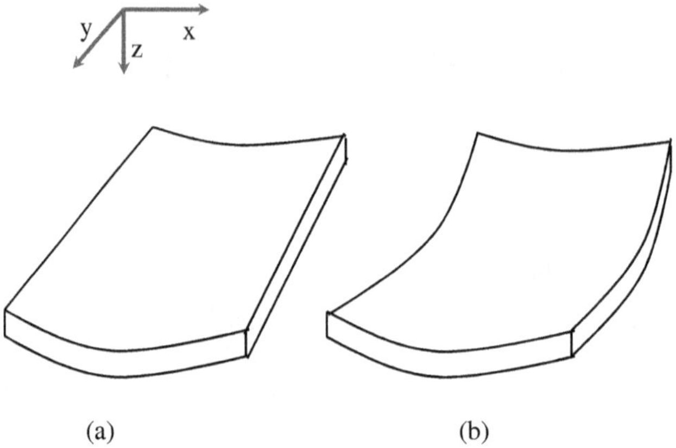

Figure 3.1 A concrete pavement slab idealized as (a) a beam or (b) a slab

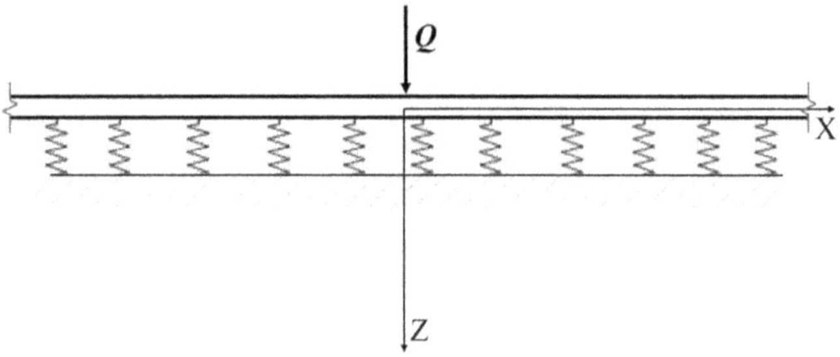

Figure 3.2 An infinite beam resting on numerous springs acted upon by a concentrated load Q

given as

$$w = e^{\lambda x}(c_1 \cos\lambda x + c_2 \sin\lambda x) + e^{-\lambda x}(c_3 \cos\lambda x + c_4 \sin\lambda x) \tag{3.4}$$

where $\lambda = \left(\frac{k}{4EI}\right)^{\frac{1}{4}}$, and c_1, c_2, c_3 and c_4 are constants. These constants can be determined from the boundary conditions pertaining to the specific geometry of the problem. Considering one side of the beam (for instance, the right side) with respect to the load application point of the infinite beam (as shown in Figure 3.2), one can write the following boundary conditions and subsequently derive the following results [119].

- $\lim_{x\to\infty} w = 0$. This condition leads to $c_1 = c_2 = 0$. Thus, the equation reduces to $w = e^{-\lambda x}(c_3 \cos\lambda x + c_4 \sin\lambda x)$
- Due to symmetry, $\frac{dw}{dx}\,|_{x=0} = 0$. This leads to $c_3 = c_4 = c$ (say)

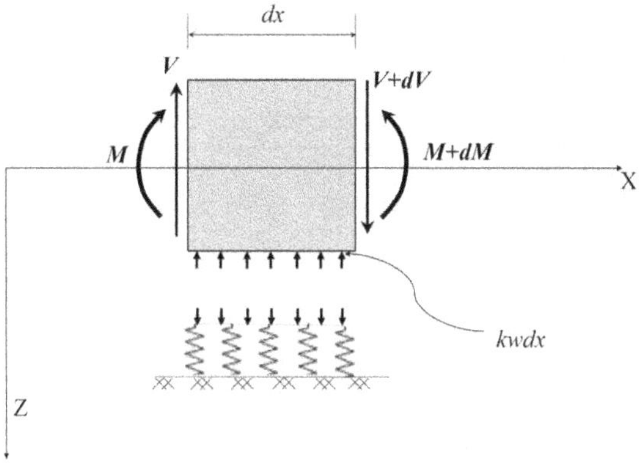

Figure 3.3 Free body diagram of dx elemental length of the beam presented in Figure 3.2

- Total upward force generated by the springs must be equal to the downward force Q applied, that is, $2\int_0^{\infty} kw\, dx = Q$. This leads to $c = \frac{Q\lambda}{2k}$

Hence, the expression for deflection of infinite beam resting on elastic foundation is obtained as

$$w = \frac{Q\lambda}{2k} e^{-\lambda x} (\cos\lambda x + \sin\lambda x) \tag{3.5}$$

It may be noted that the developed equation (Equation 3.5) is valid only for the right side (i.e. $x \geq 0$) of the infinite beam. In a similar manner, an expression can be developed for the left side of the beam. Then, the first boundary condition changes as, $\lim_{x\to-\infty} = 0$, rest two conditions remain the same. From these boundary conditions, the constants are obtained as $c_3 = c_4 = 0$, and $c_1 = -c_2 = \frac{Q\lambda}{2k}$. The equation (for the left side of the infinite beam from $x = 0$) takes the form as

$$w = \frac{Q\lambda}{2k} e^{\lambda x} (\cos\lambda x - \sin\lambda x) \tag{3.6}$$

Equations 3.5 (or 3.6) can be utilized (by successive differentiation) to obtain the rotation, bending moment and shear profile [119]. It is interesting to note that the expression for w in in Equation 3.5 and 3.6 shows positive as well as negative values at different places along x; negative values indicate that the springs are under tension, that means the sprrings are 'connected' to the beam. This is not expected to happen in a field situation. If the beam is deflected upward, the foundation support will be ineffective in those regions leading to redistribution of foundation pressure (kw) and hence w will be different than what is calculated through Equation 3.5 or 3.6.

The maximum deflection is under the load and is obtained as

$$w_{\text{max}} = \frac{Q\lambda}{2k} \tag{3.7}$$

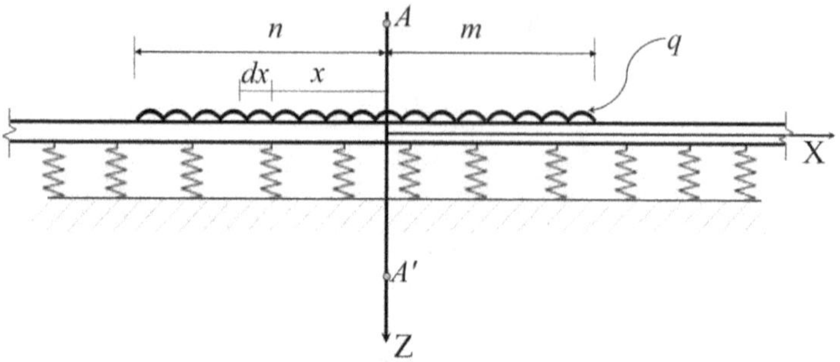

Figure 3.4 Analysis of infinite beam with distributed loading

In case, there is a uniformly distributed loading (instead of point loading) of q per unit length, the deflection can be obtained by integration.

For the loading diagram shown as in Figure 3.4, the deflection at AA' ($w_{AA'}$) can be obtained by using superposition of deflection calculated from Equations 3.5 and 3.6 and can be expressed as follows:

$$w_{AA'} = \int_0^n \frac{q\lambda}{2k} e^{-\lambda x} (\cos\lambda x + \sin\lambda x)\,dx + \int_{-m}^0 \frac{q\lambda}{2k} e^{\lambda x} (\cos\lambda x - \sin\lambda x)\,dx \tag{3.8}$$

If the section AA' is outside the loaded area (of length $n+m$), Equations 3.5 or 3.6 need to be used with appropriate integration limits. Further, a beam can be semi-infinite (that is, the beam has a definite ending at one side, and the other side is infinite), or finite (that, is the beam has a finite length). For example, finite beam can be considered as a closer approximation of concrete slab than semi-infinite and infinite beam idealizations.

Example problem

Refer to Figure 3.5. A semi-infinite beam is acted upon by a concentrated load Q at its free edge. If the general solution for the displacement is given as Equation 3.4, find out the expression for its deflection.

The choice of the coordinate is shown in the Figure 3.5 itself. The boundary conditions are,

Solution

Considering $\lim_{x\to\infty} w = 0$, $c_1 = c_2 = 0$. Thus, the equation reduces to $w = e^{-\lambda x}(c_3 \cos\lambda x + c_4 \sin\lambda x)$

Considering moment at the free-edge is zero, $EI\frac{d^2w}{dx^2}\,|_{x=0} = 0$; this leads to $c_4 = 0$.

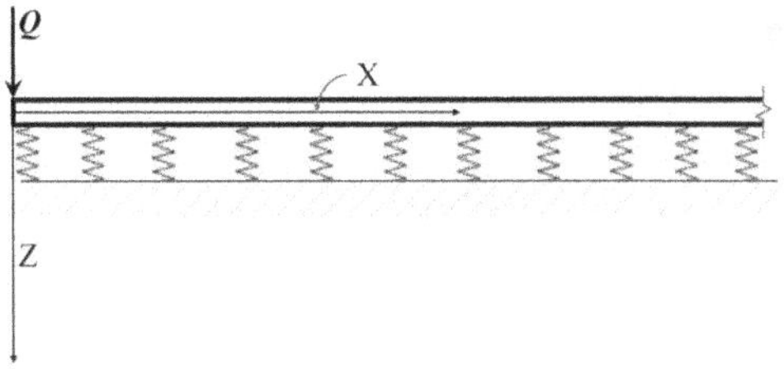

Figure 3.5 A semi-infinite beam resting on numerous springs acted upon by a concentrated load Q at its free end

Considering Q load is applied at the free edge, the magnitude of the shear force at $x = 0$ will also be equal to it. Hence, $-EI\frac{d^3w}{dx^3}|_{x=0} = -Q$; this leads to $c_3 = \frac{2Q\lambda}{k}$

Thus, the expression for deflection will be

$$w = \frac{2Q\lambda}{k}e^{-\lambda x}\cos\lambda x \tag{3.9}$$

The solution of the above problem becomes more involved in case the concentrated load is applied somewhere away to the right side from the free edge. One of the ways to solve the problem could be deal beam in multiple parts and use compatibility conditions between these parts. Alternatively, one can solve such problems as a superposition of two infinite beams. This technique can be used for solving semi-infinite or finite beam acted upon by concentrated load, distributed load or moment positioned anywhere on the beam. One can refer to [119, 258], for example, for the details this superposition approach.

Such 1-D analysis is useful, for example, for analysis problem of dowel bar. Figure 3.6 illustrates how a single dowel bar can be idealized as a finite beam resting on elastic foundation. However, additional considerations are involved in dowel bar analysis problem (refer Figure 3.6), for example, (i) there is a discontinuity of support in the middle portion, (ii) one side of the dowel bar is embedded in concrete but the other side is free to move horizontally, (iii) the wheel load does not directly act on dowel bar, etc. Interested readers can refer to past works by Friberg [92, 93] and Bradbury [35] and relatively recent study by Porter [231] on dowel bar analysis and the assumptions involved.

In line with the development of Equation 3.3, a equilibrium condition of a beam (refer Equation 3.1) with an arbitrary loading (of q per unit length, which may include self-weight) and arbitrary foundation support (of p per unit length) condition (refer Figure 3.7) can be written as

$$\frac{dV}{dx} = p - q \tag{3.10}$$

Or,

$$EI\frac{d^4w}{dx^4} = q - p = q^* \tag{3.11}$$

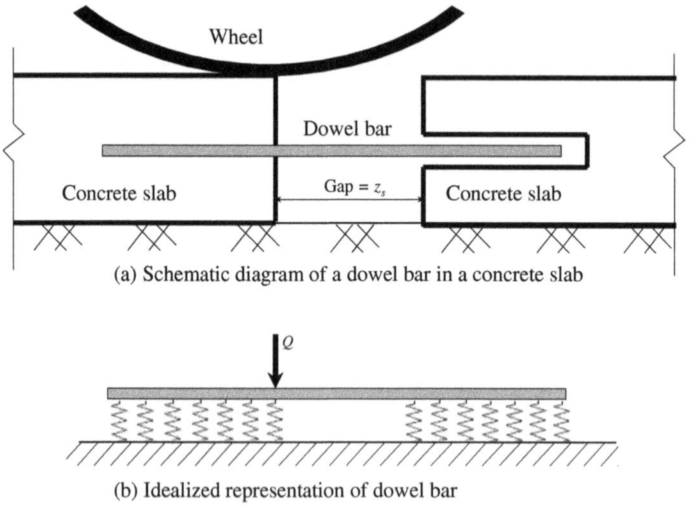

Figure 3.6 Idealization of dowel bar for analysis

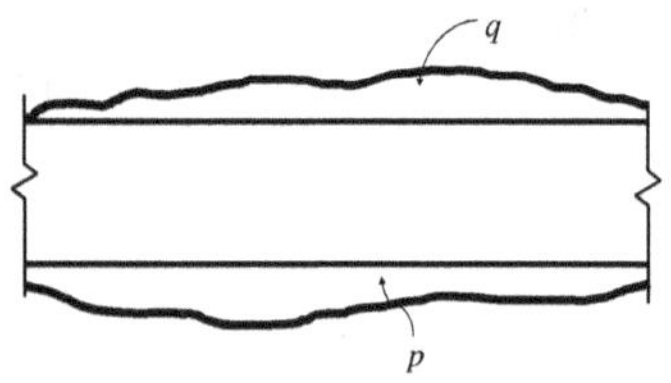

Figure 3.7 A beam with arbitrary loading

where q^* is net loading per unit length in the downward direction. Depending on the foundation support, the expression for p may become different. It may be noted that if $p = 0$, it becomes equivalent to beam bending equation, without any spring support. Winkler, Pasternak, Kerr are the examples of different types of supports and are briefly discussed in the following.

3.2.1 BEAM RESTING ON WINKLER FOUNDATION

Mutually independent (linear) springs are known as Winkler's spring [130, 146, 148, 189]. Formulation for beam resting on Winkler spring for a pointed loading has been already discussed in the beginning of this section (Section 3.2). That is, for Winkler spring, $p = kw$. Thus, a beam resting on Winkler's spring subjected to loading q (following Equation 3.3) can be represented as

$$EI\frac{d^4w}{dx^4} = q - kw \tag{3.12}$$

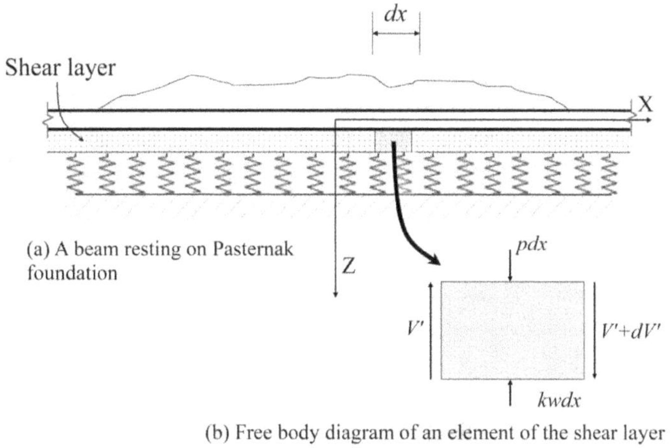

Figure 3.8 A beam resting on a Pasternak foundation and a free-body-diagram of the shear layer

The Winkler's spring constant (k) used here in the formulation indicates the pressure needed on the spring system to cause unit displacement[2]. It's unit is therefore MPa/mm. As discussed in Section 2.2, modulus of subgrade reaction (k) is also pressure needed to cause unit deformation to the medium (that is, subgrade or sub-base or base layer). Thus, the spring constant used in the present formulation is conceptually equivalent as modulus of subgrade reaction of the supporting layer.

Modulus of subgrade reaction seems to be the function of the dimension of the foundation[3], and hence non-unique. Discrepancies in the deflection values have been observed between the field results and model predictions (especially at the edges and corners of the slab), these idealized springs being mutually independent members. One can, for example, refer to [65, 127, 239, 282] for discussions on the issues involved. This has prompted researchers to develop multi-parameter models as an improvements to over the Winkler's foundation model so to capture the response of the foundation support (subgrade) better. Some of these are discussed in the following.

3.2.1.1 Beam Resting on Pasternak Foundation

In a Pasternak foundation, it is assumed that there is a hypothetical shear layer placed at the top of the spring system (refer Figure 3.8(a)). Thus, considering the equilibrium of an element of length dx of shear layer (refer Figure 3.8(b)), one can write

$$pdx - V' + (V' + dV') - kwdx = 0$$

$$p = kw - \frac{dV'}{dx} \tag{3.13}$$

[2]Winkler model is also known as dense liquid model, and k represents the pressure needed to cause unit vertical displacement to a hypothetical floating body against buoyancy.

[3]The larger is the size of foundation, the larger is the portion of soil that gets mobilized.

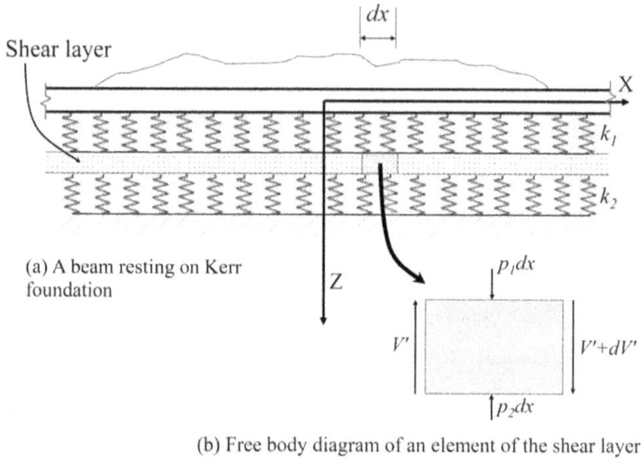

Figure 3.9 A beam resting on a Kerr foundation and a free-body-diagram of the shear layer

If the shear force developed within the shear layer is assumed to be proportional to the slope, then it can be written

$$V' = G_s \frac{dw}{dx} \tag{3.14}$$

where G_s is the shear modulus of the foundation. Putting, Equation 3.14 in Equation 3.13, one obtains

$$p = kw - G_s \frac{d^2w}{dx^2} \tag{3.15}$$

Putting, Equation 3.15 in Equation 3.11 (i.e. the general equation for beam resting on elastic foundation), the equation for 1-D beam resting on Pasternak foundation becomes

$$EI \frac{d^4w}{dx^4} - G_s \frac{d^2w}{dx^2} + kw = q \tag{3.16}$$

One can refer to, for example, [45] for worked out solutions for various problem geometries on Pasternak foundation.

3.2.2 BEAM RESTING ON KERR FOUNDATION

Kerr foundation model [148, 149] consists of two layers of Winkler springs (with spring constants k_1 and k_2, say) with a shear layer in between (refer Figure 3.9(a)). The free body diagram of an element of length dx of the shear layer in the Kerr foundation is shown in Figure 3.9(b). The pressures transmitted on the top and the bottom of the shear layer are shown as p_1 and p_2, and the displacements the top and the bottom set of springs undergo are w_1 and w_2, respectively.

Considering the equilibrium of the shear layer,

$$\begin{aligned} p_1 - p_2 &= -\frac{dV'}{dx} \\ &= -G_s \frac{d^2 w_2}{dx^2} \text{ (Similar to Equation 3.14)} \end{aligned} \quad (3.17)$$

Further,

$$w = w_1 + w_2 \quad (3.18)$$

where w is the deflection of the beam. The spring conditions can be written as

$$\begin{aligned} p_1 &= k_1 w_1 \\ p_2 &= k_2 w_2 \end{aligned}$$

Using the above in Equation 3.17, one can write

$$\begin{aligned} p_1 &= k_2 w_2 - G\frac{d^2 w_2}{dx^2} \\ &= k_2\left(w - \frac{p_1}{k_1}\right) - G\frac{d^2}{dx^2}\left(w - \frac{p_1}{k_1}\right) \end{aligned} \quad (3.19)$$

Putting Equation 3.19 in the Equation 3.11 (i.e. the general equation for beam resting on elastic foundation and considering that $p = p_1$ in the present case), it can be written as

$$\begin{aligned} \frac{G_s EI}{k_1}\frac{d^6 w}{dx^6} - \left(1 + \frac{k_2}{k_1}\right) EI\frac{d^4 w}{dx^4} + G_s\frac{d^2 w}{dx^2} - k_2 w \\ = \frac{G}{k_1}\frac{d^2 q}{dx^2} - \left(1 + \frac{k_2}{k_1}\right) q \end{aligned} \quad (3.20)$$

3.2.3 VARIOUS OTHER MODELS

Figure 3.10 shows free body diagram of an element of a beam, which is resting on an elastic foundation and is also subjected to an axial force.

Taking moment with respect to point 'O' and neglecting the higher order terms, one obtains [119]

$$\begin{aligned} dM - V_d dx - F(-dw) &= 0 \\ \text{Or, } \frac{dM}{dx} + F\frac{dw}{dx} - V_d &= 0 \\ \text{Or, } \frac{d^2 M}{dx^2} + F\frac{d^2 w}{dx^2} - \frac{dV_d}{dx} &= 0 \end{aligned} \quad (3.21)$$

It may be noted that oblique element has been taken here (hence shear force has been denoted as V_d instead of V). However, it can be shown that even if an rectangular

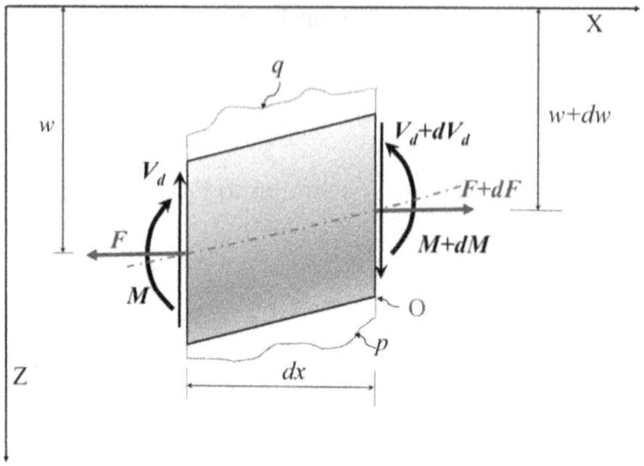

Figure 3.10 A beam resting on elastic foundation is also subjected axial force

element is taken (with sides normal to the deflected shape), the same relationship can be reached [287].

Considering $EI\frac{d^2w}{dx^2} = -M$ and Equation 3.10 one can write

$$EI\frac{d^4w}{dx^4} - F\frac{d^2w}{dx^2} + p = q \tag{3.22}$$

It can be seen that the form of Equation 3.22 is similar to Equation 3.16 (hence the solution approach will be similar), even though these are derived for two different types of problems. Once can refer to [119] for the detailed solution.

Models can be further developed to represent geotextile/geogrid placed within the shear layer [186, 188]. There can be situations where, friction between granular/soil material with the geotextile/geogrid is mobilized, and the tension varies along x. The spring foundation model also has been applied to represent other field situations, for example support from non-homogeneous soil [203].

There are a large number of various such models of beams on elastic foundation and varieties of solution techniques proposed and studied by various researchers [77, 244]. For example, the Filonenko-Borodich model assumes a thin elastic membrane under constant tension is connecting the springs, the Heténi model assumes a beam or plate with known flexural rigidity is connected with the springs, the Rhines model assumes that the shear layer (used in Pasternak model or Kerr model) shows elasto-plastic response and so on. Interested readers may refer to, for example papers/reports such as [130, 146, 148, 159] for a review on various types of foundation models or refer to the book by Selvadurai [258] for a detailed discussion.

Other than the spring models (generally classified as 'lumped parameter model'), continuum models are also used to represent subgrade support. This has been discussed in Chapter 5.

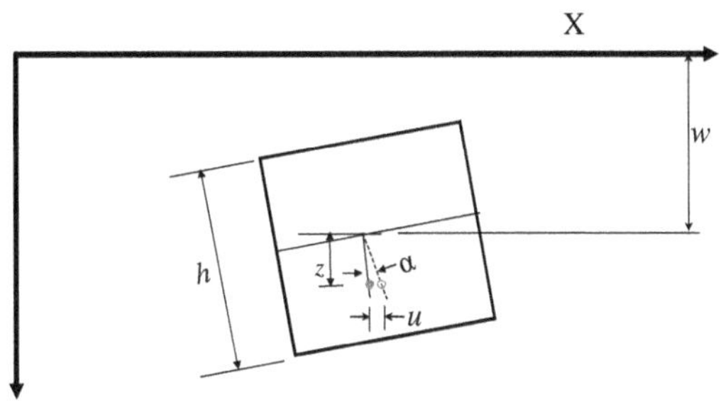

Figure 3.11 Deflected position of an element of a thin plate

3.3 ANALYSIS OF THIN PLATE RESTING ON ELASTIC FOUNDATION

Concrete slabs are generally idealized as thin plate resting on elastic foundation. In the following, the theory of thin plates resting on elastic foundation (Winkler, Pasternak or Kerr foundation) has been presented. The principles and formulations of plate theories are widely used in practice and research. One can refer to, for example, [107, 139, 258, 286, 309] for detailed discussions. Here, the discussion on theory of plate has been kept brief, and interested readers may refer to the above books, for instance, for further study.

The assumptions for a thin plate with small deflection are as follows [107, 139, 159, 258, 286]:

- The thickness of the plate (h) is small compared to its other dimensions.
- The deflection w is small compared to the thickness of the plate.
- Middle plane does not get stretched due to application of load.
- Plane sections perpendicular to the middle plane remains plane before and after bending.
- The plate can deform in only two ways; it can expand or contract axially, or it can bend with its cross section remaining plane. Therefore, normal stresses along transverse direction can be ignored i.e., $\sigma_{zz} = 0$.

3.3.1 PLATE RESTING ON WINKLER'S FOUNDATION

The thickness of the plate is h, and its middle plane assumed to coincide with the X–Y plane; therefore, z varies from $-\frac{h}{2}$ to $\frac{h}{2}$. The deflection of the plate is assumed as u and v in the directions of X and Y, respectively. Figure 3.11 shows an element of the deflected plate, and deflection is shown along the X direction only. From Figure 3.11, one can write

$$\begin{aligned} u &= -z\sin\alpha \\ &\approx -z\tan\alpha \\ &\approx -z\frac{\delta w}{\delta x} \end{aligned} \tag{3.23}$$

Similarly, there will deflection along the Y direction, hence

$$v \approx -z\frac{\delta w}{\delta y} \tag{3.24}$$

Using Equations 3.23 and 3.24 in Equation 1.18, one can write

$$\begin{aligned}
\varepsilon_{xx} &= -z\frac{\delta^2 w}{\delta x^2} \\
\varepsilon_{yy} &= -z\frac{\delta^2 w}{\delta y^2} \\
\gamma_{xy} &= -2z\frac{\delta^2 w}{\delta x \delta y}
\end{aligned} \tag{3.25}$$

Substituting the expressions in Equation 3.25 in Equation 1.34 (that is, for plane stress conditions), one obtains

$$\begin{aligned}
\sigma_{xx} &= -\frac{Ez}{1-\nu^2}\left(\frac{\delta^2 w}{\delta x^2} + \nu\frac{\delta^2 w}{\delta y^2}\right) \\
\sigma_{yy} &= \frac{E}{1-\nu^2}(\varepsilon_{yy} + \nu\varepsilon_{xx}) = -\frac{Ez}{1-\nu^2}\left(\frac{\delta^2 w}{\delta y^2} + \nu\frac{\delta^2 w}{\delta x^2}\right) \\
\tau_{xy} &= \frac{E\gamma_{xy}}{2(1+\nu)} = -\frac{Ez}{2(1+\nu)}\frac{\delta^2 w}{\delta x \delta y}
\end{aligned} \tag{3.26}$$

Using Equation 3.26, the moments are obtained as

$$\begin{aligned}
M_{xx} &= \int_{-\frac{h}{2}}^{\frac{h}{2}} \sigma_{xx} z dz \\
&= -\frac{Eh^3}{12(1-\nu^2)}\left(\frac{\delta^2 w}{\delta x^2} + \nu\frac{\delta^2 w}{\delta y^2}\right) = -D\left[\frac{\delta^2 w}{\delta x^2} + \nu\frac{\delta^2 w}{\delta y^2}\right]
\end{aligned} \tag{3.27}$$

where $D = \frac{Eh^3}{12(1-\nu^2)}$ is the flexural rigidity of a plate.

$$\begin{aligned}
M_{yy} &= \int_{-\frac{h}{2}}^{\frac{h}{2}} \sigma_{yy} z dz \\
&= -\frac{Eh^3}{12(1-\nu^2)}\left(\frac{\delta^2 w}{\delta y^2} + \nu\frac{\delta^2 w}{\delta x^2}\right) = -D\left[\frac{\delta^2 w}{\delta y^2} + \nu\frac{\delta^2 w}{\delta x^2}\right]
\end{aligned} \tag{3.28}$$

$$\begin{aligned}
M_{xy} &= -\int_{-\frac{h}{2}}^{\frac{h}{2}} \tau_{xy} z dz \\
&= \frac{Eh^3}{12(1+\nu)}\frac{\delta^2 w}{\delta x \delta y} = D(1-\nu)\frac{\delta^2 w}{\delta x \delta y}
\end{aligned} \tag{3.29}$$

Equations 3.27 to 3.29 are the widely used moment equations in thin plate theories. It may be recalled that in the present case the thin plate is assumed to be resting on elastic foundation. Let the net vertical downward pressure be assumed as q^* per unit area (that is $q^* = q - p$), where q is the applied load per unit area, which may include the self-weight of the slab, and p is the foundation support, per unit area, provided by the springs (for example, Winkler, Pasternak, Kerr, etc.) per unit area.

Figure 3.12 shows the various forces and moments acting on a small element $(dx \times dy)$ of the plate. The assumed positive directions are also shown on the diagram. For the purpose of clarity, the diagram has been divided in three parts (a), (b) and (c). It may be noted that the moments and shear forces are acting per unit length, whereas q^* is acting per unit area. By taking vertical force equilibrium (refer Figure 3.12(a)), one obtains

$$\frac{\delta V_{xx}}{\delta x}dxdy + \frac{\delta V_{yy}}{\delta y}dxdy + q^* dxdy = 0$$

$$\text{Or, } \frac{\delta V_{xx}}{\delta x} + \frac{\delta V_{yy}}{\delta y} = -q^* \tag{3.30}$$

By taking moment along DC, one obtains

$$-\frac{\delta M_{yy}}{\delta y}dxdy + \frac{\delta M_{xy}}{\delta x}dxdy + V_{yy}dxdy - q^* dxdy\frac{dy}{2} - \frac{\delta V_{xx}}{dx}dxdy\frac{dy}{2} = 0 \tag{3.31}$$

Neglecting smaller order terms and rearranging,

$$\frac{\delta M_{xy}}{\delta x} - \frac{\delta M_{yy}}{\delta y} + V_{yy} = 0 \tag{3.32}$$

Similarly, by taking moment along BC, one obtains

$$\frac{\delta M_{xx}}{\delta x}dxdy + \frac{\delta M_{yx}}{\delta y}dxdy - V_{xx}dxdy - q^* dxdy\frac{dx}{2} - \frac{\delta V_{yy}}{dx}dxdy\frac{dx}{2} = 0 \tag{3.33}$$

Similarly, neglecting smaller order terms and rearranging,

$$\frac{\delta M_{xx}}{\delta x} + \frac{\delta M_{yx}}{\delta y} - V_{xx} = 0 \tag{3.34}$$

Putting Equations 3.32 and 3.34 in Equation 3.30, one obtains [107,139,258,286]

$$\frac{\delta^2 M_{yy}}{\delta y^2} - \frac{\delta^2 M_{xy}}{\delta x \delta y} + \frac{\delta^2 M_{xx}}{\delta x^2} + \frac{\delta^2 M_{yx}}{\delta x \delta y} + q^* = 0$$

$$\text{Or, } \frac{\delta^2 M_{xx}}{\delta x^2} - 2\frac{\delta^2 M_{xy}}{\delta x \delta y} + \frac{\delta^2 M_{yy}}{\delta y^2} + q^* = 0 \tag{3.35}$$

since, $M_{xy} = -M_{yx}$.

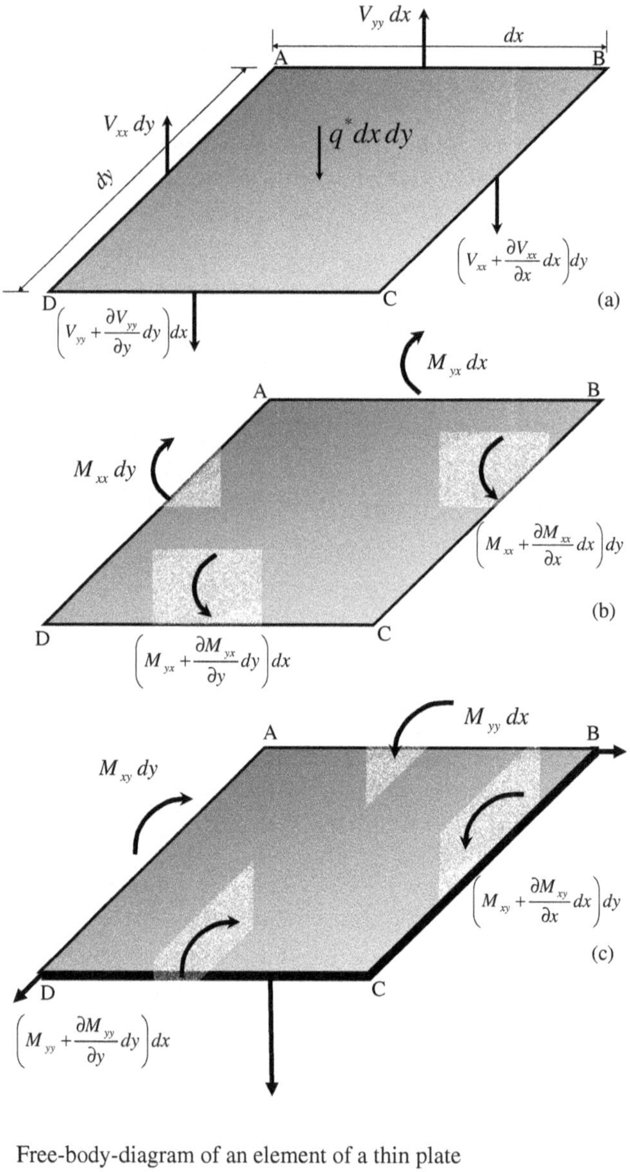

Figure 3.12 Free-body-diagram of an element of a thin plate

Now, putting expressions for M_{xx}, M_{yy} and M_{xy} from Equations 3.27, 3.28 and 3.29, respectively, one obtains [107, 139, 258, 286]

$$-D\frac{\delta^2}{\delta x^2}\left[\frac{\delta^2 w}{\delta x^2}+\nu\frac{\delta^2 w}{\delta y^2}\right]-2D(1-\nu)\frac{\delta^4 w}{\delta x^2\delta y^2}$$
$$-D\frac{\delta^2}{\delta y^2}\left[\frac{\delta^2 w}{\delta y^2}+\nu\frac{\delta^2 w}{\delta x^2}\right]+q^*=0$$

$$\text{Or,}\ -D\left[\frac{\delta^4 w}{\delta x^4}+2\frac{\delta^4 w}{\delta x^2\delta y^2}+\frac{\delta^4 w}{\delta y^4}\right]+q^*=0$$

$$\text{Or,}\ D\nabla^4 w=q^* \tag{3.36}$$

where $\nabla^2=\frac{\delta^2}{\delta x^2}+\frac{\delta^2}{\delta y^2}$

Equation 3.36 is a well-known equation for plate resting on spring foundation [107, 139, 258, 286, 309]. For the Winkler's foundation case, the equation can be re-arranged as

$$D\nabla^4 w+kw=0 \tag{3.37}$$

The equation can as well be written as

$$l^4\nabla^4 w+w=0 \tag{3.38}$$

where $l=\left(\frac{Eh^3}{12(1-\nu^2)k}\right)^{1/4}$. l is also known as radius of relative stiffness [131, 318].

Further, the cases of a plate resting on Pasternak and Kerr foundations are discussed briefly in the following sections. Proceeding in the similar manner in cylindrical coordinate system also one can show that the Equation 3.36 holds, where, $\nabla^2=\frac{\partial^2}{\partial r^2}+\frac{1}{r}\frac{\partial}{\partial r}+\frac{1}{r^2}\frac{\partial^2}{\partial\theta^2}$. One can, for example, refer to [139, 286] for the derivation.

3.3.2 PLATE RESTING ON PASTERNAK FOUNDATION

The free body diagram of the shear layer for a 2-D Pasternak foundation is shown in Figure 3.13. Considering the force equilibrium of the shear layer, one can write

$$pdxdy+\frac{\partial V_{xx}^{'}}{\partial x}dxdy+\frac{\partial V_{yy}^{'}}{\partial y}dxdy-kwdxdy=0 \tag{3.39}$$

Assuming that shear forces developed in the shear layer is proportional to the slope (in a similar manner as done in Section 3.2.1.1), one can be write

$$V_{xx}^{'}=G\frac{\partial w}{\partial x},\ \text{and}\ V_{yy}^{'}=G\frac{\partial w}{\partial y} \tag{3.40}$$

Thus, one can be write

$$p=kw-G\nabla^2 w \tag{3.41}$$

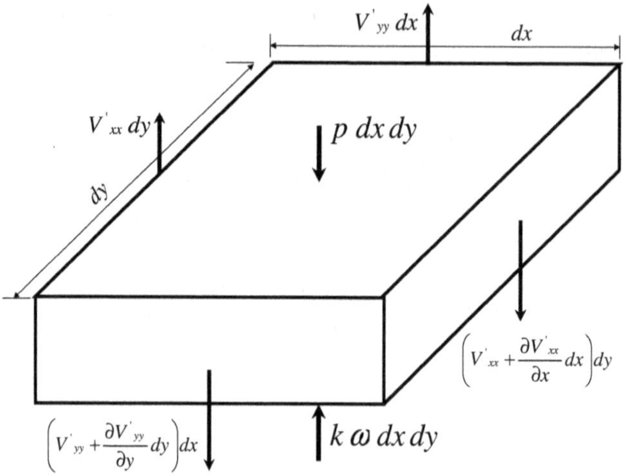

Figure 3.13 Free-body-diagram of the shear layer of a 2-D Pasternak foundation

Therefore, considering Equation 3.36, the governing equation for a plate resting on Pasternak foundation become

$$D\nabla^4 w = q^* = q - p = \quad q - (kw - G\nabla^2 w)$$

$$\text{Or, } D\nabla^4 w - G\nabla^2 w + kw = q \tag{3.42}$$

3.3.3 PLATE RESTING ON KERR FOUNDATION

Proceeding in the similar manner as presented in Section 3.2.2, for 2-D case, one can obtain the governing equation for Kerr foundation as

$$\frac{GD}{k_1}\nabla^6 w - \left(1+\frac{k_2}{k_1}\right)D\nabla^4 w + G\nabla^2 w - k_2 w = \frac{G}{k_1}\nabla^2 q - \left(1+\frac{k_2}{k_1}\right)q \tag{3.43}$$

One can refer to the paper by Cauwelaert et al. [47] where a detailed development of the formulation and solution for Kerr foundation case has been provided.

3.3.4 BOUNDARY CONDITIONS

Figure 3.14 shows a concrete pavement slab of dimension $L \times B$, idealized as a thin plate. For solving the plate equation (that is, Equation 3.36), suitable boundary conditions (of the individual four edges) need to be incorporated. The following boundary conditions are possible for any of the edges (say, for the edge $y = L$) [139, 286]

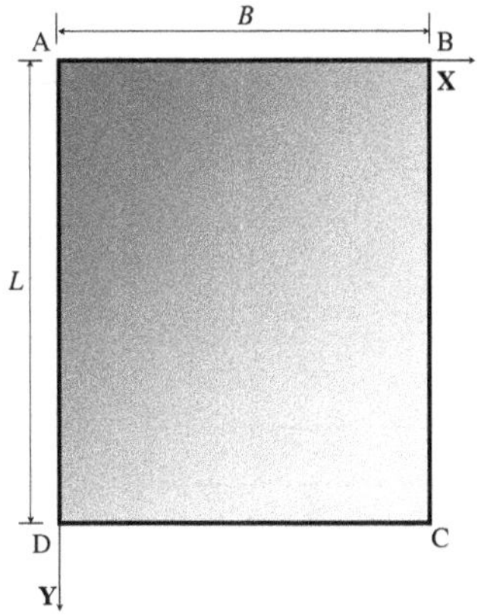

Figure 3.14 The slab boundaries may be free, fixed or hinged

- The edge $y = L$ may be fixed. Then, the deflection and the slope will be equal to zero. That is, $w\,|_{y=L} = 0$ and $\left(\frac{\delta w}{\delta y}\right)|_{y=L} = 0$
- The edge $y = L$ may be hinged. Then, the deflection and the moment along that direction will be equal to zero. That is, $w\,|_{y=L} = 0$ and $M_{yy} = 0$. The second condition implies, $-D\left(\frac{\delta^2 w}{\delta y^2} + \nu\frac{\delta^2 w}{\delta x^2}\right)|_{y=L} = 0$. Further, it may be noted that since there will be no deflection along $y = L$ line, $\left(\frac{\delta^2 w}{\delta x^2}\right)_{y=L}$ will be equal to zero. This will finally result the second condition as, $\frac{\delta^2 w}{\delta y^2}\,|_{y=L} = 0$
- The edge $y = L$ may be free. In that case, moments will be zero. That is, $M_{yy} = 0$ and $M_{yx} = 0$.

However, for a concrete pavement, the edges are not exactly free, fixed or hinged. Presence of dowel and tie bars (along the longitudinal and transverse direction respectively) makes the edge conditions somewhere in-between. Although researchers have addressed such issues [178,336], a simple situation with all the edges as simply supported is taken up in the next section.

3.4 LOAD STRESS IN CONCRETE PAVEMENT SLAB

If a concrete pavement slab can be idealized as a thin plate resting on Winkler's foundation, the solution of Equation 3.36 will provide an estimate of the stresses due to load. The dimension of the plate is assumed as $L \times B$, and all edges are considered as simply supported. It is assumed that a uniform loading of magnitude q_o per unit area is acting on a rectangular area of dimension $l \times b$. The centre of this area is located at $(\bar{x}, \bar{y})$. The arrangement is schematically presented in Figure 3.15.

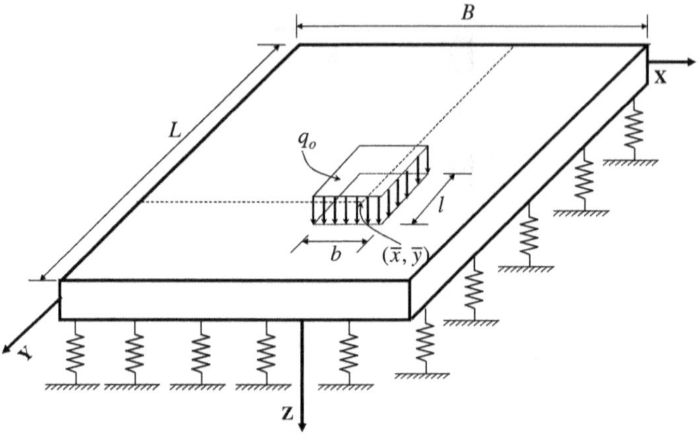

Figure 3.15 Slab resting on Winkler's spring acted upon by a rectangular patch loading

Let, a loading expressed in the form of Equation 3.44 is applied[4] to the plate [107, 139, 286]

$$q = q_o \sin\frac{m\pi x}{L} \sin\frac{n\pi y}{B} \tag{3.44}$$

where m and n are any numbers.

Then, Equation 3.44 is a possible solution to Equation 3.36. The boundary conditions (Refer to Section 3.3.4 for a discussion on boundary conditions) are satisfied if w is expressed as [107, 139, 286, 309],

$$w = c \sin\frac{m\pi x}{L} \sin\frac{n\pi y}{B} \tag{3.45}$$

where c is a constant. The value of c can be obtained by putting Equation 3.44 and 3.45 in Equation 3.36, as [139, 286]

$$c = \frac{q_o}{\left(\pi^4 D\left(\frac{m^2}{L^2} + \frac{n^2}{B^2}\right)^2 + k\right)} \tag{3.46}$$

Thus, the expression for deflection w becomes

$$w = \frac{q_o}{\left(\pi^4 D\left(\frac{m^2}{L^2} + \frac{n^2}{B^2}\right)^2 + k\right)} \sin\frac{m\pi x}{L} \sin\frac{n\pi y}{B} \tag{3.47}$$

Equation 3.47 provides a solution for single sinusoidal loading (Equation 3.44). It is possible to express any loading function $q = f(x,y)$ (acting over a given area),

[4]That is, the loading is assumed to be different than uniformly distributed loading on rectangular area shown in Figure 3.15.

as the sum of series of sinusoidal loadings [107, 139, 286], as per Navier's transformation. That is,

$$\begin{aligned} q &= f(x,y) \qquad (3.48) \\ &= \sum_{m=1}^{\infty}\sum_{n=1}^{\infty} c_{mn} \sin\frac{m\pi x}{L} \sin\frac{n\pi y}{B} \end{aligned}$$

where

$$c_{mn} = \frac{4}{LB}\int_{\text{Area}} f(x,y) \sin\frac{m\pi x}{L} \sin\frac{n\pi y}{B} dxdy \qquad (3.49)$$

The expression for w in the present case, therefore, becomes [107, 139, 286],

$$w = \sum_{m=1}^{\infty}\sum_{n=1}^{\infty} \frac{c_{mn}}{\left(\pi^4 D\left(\frac{m^2}{L^2}+\frac{n^2}{B^2}\right)^2 + k\right)} \sin\frac{m\pi x}{L} \sin\frac{n\pi y}{B} \qquad (3.50)$$

In the present case, the load is uniformly distributed over a rectangular area of $l \times b$, centre of which is located at $(\bar{x}, \bar{y})$. That is, $q = f(x,y) = \frac{Q}{lb}$ within the region bound between $x = \bar{x} - \frac{b}{2}$ and $\bar{x} + \frac{b}{2}$ and $y = \bar{y} - \frac{l}{2}$ and $\bar{y} + \frac{l}{2}$ (refer Figure 3.15). The value of c_{mn} is calculated as [107, 139, 286]:

$$\begin{aligned} c_{mn} &= \frac{4Q}{LBlb}\int_{\bar{x}-\frac{b}{2}}^{\bar{x}+\frac{b}{2}}\int_{\bar{y}-\frac{l}{2}}^{\bar{y}+\frac{l}{2}} \sin\frac{m\pi x}{L} \sin\frac{n\pi y}{B} dxdy \\ &= \frac{16Q}{\pi^2 LBlbmn} \sin\frac{m\pi\bar{x}}{L} \sin\frac{n\pi\bar{y}}{B} \sin\frac{m\pi l}{2L} \sin\frac{n\pi b}{2B} \qquad (3.51) \end{aligned}$$

The expression for w is therefore [107, 139, 286]

$$\begin{aligned} w = \frac{16Q}{\pi^2 LBlb} \sum_{m=1}^{\infty}\sum_{n=1}^{\infty} \frac{1}{mn\left(\pi^4 D\left(\frac{m^2}{a^2}+\frac{n^2}{b^2}\right)^2 + k\right)} \sin\frac{m\pi\bar{x}}{L} \\ \sin\frac{n\pi\bar{y}}{B} \sin\frac{m\pi l}{2L} \sin\frac{n\pi b}{2B} \sin\frac{m\pi x}{L} \sin\frac{n\pi y}{B} \qquad (3.52) \end{aligned}$$

Once the expression for w is obtained (from Equation 3.52), this can be used to find out bending moment (refer Equations 3.27, 3.28 and 3.29) or the bending stress (refer Equation 3.26). This is one of the possible approaches of obtaining the solution of Equation 3.36, and one may refer to, for example, [258, 286, 309] for alternative approaches and numerical methods of solution.

Figure 3.16 presents a schematic diagram showing the variation of maximum bending stress with slab thickness (h) and modulus of subgrade reaction (k) (these terms appear in the expressions of D and q^*, respectively, in Equation 3.36). It may be noted that the maximum bending stress will always occur below the wheel(s). Figure 3.16 shows that the maximum bending stress (say at interior) decreases with the (i) increase in the slab thickness and/or (ii) increase in the modulus of subgrade reaction.

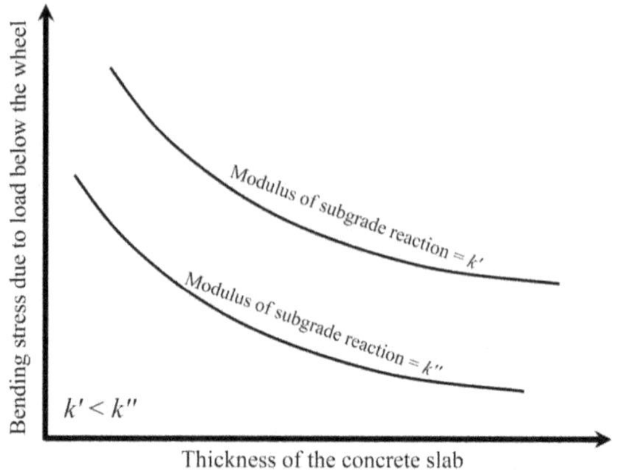

Figure 3.16 Schematic diagram showing variation of bending stress (due to load at interior) with slab thickness and modulus of subgrade reaction

In the above formulation, the springs can take tension as well; however, in reality underlaying layer does not pull back the concrete slab when it tries to bend upwards. Thus, tension, if arises in the analysis, should be made equal to zero; yet, the portion (in which tension would arise) is not known a-priory. Thus, the solution needs to be obtained iteratively, and achieving a closed-form solution may become difficult (similar discussion has been made in Section 3.2).

Further, it is assumed that the rectangular loaded area approximately represents the tyre imprint. However, actual tyre imprint may not necessarily be rectangular in shape and nor be uniform in loading (refer to Figure 5.9).

3.5 CLOSURE

Individual solutions has been obtained for edge, interior and corner loading and varied boundary and loading conditions using the basic Equation 3.36. Pioneering works were done by Westergaard through his publications ranging from 1923 to 1946 [317–320]. Interested readers can refer to [130] for the background and historical details on the analysis of concrete pavements.

Researchers have provided solutions for thick and thin plates with various foundations, for example, thick plate resting on Pasternak foundation [266, 267], thick plate resting on Winkler's foundation [97], thin plate resting on Kerr foundation [47] and various other boundary and foundation conditions [131, 151].

A number of softwares/algorithms have been developed which can perform analyses of concrete pavements, for example [58, 72, 128, 130, 136, 220, 279]. Formulas and analysis charts/tables are also available in books/codes/guidelines for estimation of load stress in concrete pavement for various given loading configuration, base/sub-base strength and trial slab thickness [1, 136, 214, 220, 290, 292].

4 Temperature Stress in Concrete Pavement

4.1 INTRODUCTION

A pavement can be considered thermally stress-free at a given temperature (in fact, it is not a single value of a temperature, but a thermal profile as discussed later in Section 4.3). Any temperature other than the zero-stress temperature will induce thermal strains to the pavement. If the pavement is free to move, no thermal stress will be generated. But restraint (full or partial) provided by the self-weight of the concrete slab, and/or the friction between the concrete slab and the underlying layer may prevent free movement of the slab, and hence thermal stress will be generated. These issues have been covered in this chapter.

4.2 THERMAL PROFILE

The top surface of the pavement and layers below does not have the same temperature. Thus, a thermal gradient exists across the pavement thickness. The top surface of the pavement is exposed to the weather. Hence, the pavement surface temperature undergoes variations with the variation of the weather conditions. The angle of incidence of sun-rays, sky cover, wind speed, humidity, rain, etc., affects the pavement surface temperature [109, 118, 143, 315]. In the following, a one-dimensional formulation has been presented for estimation of the thermal profile across the depth of the pavement, under steady-state condition.

Consider an element[1] of unit thickness on the X–Z plane, as shown in Figure 4.1. The diagram shows that heat of Q_h per unit area is entering the element and heat of $Q_h + dQ_h$ per unit area is leaving the element due to temperature difference of dT between the thickness of the element dz. Thus, the heat balance equation can be written as

$$Q_h dxdt - (Q_h + dQ_h)dxdt = dx\,dz\,\rho\,C^h\,dT \tag{4.1}$$

where ρ is the density of material and C^h is the heat capacity. Thus,

$$-\frac{dQ_h}{dz} = \rho C^h \frac{dT}{dt} \tag{4.2}$$

As per Fourier's law of heat conduction, one can write

$$Q_h = -k^{td} \rho C^h \frac{dT}{dz} \tag{4.3}$$

[1]homogenous, isotropic and constant thermal properties

DOI: 10.1201/9781003190769-4

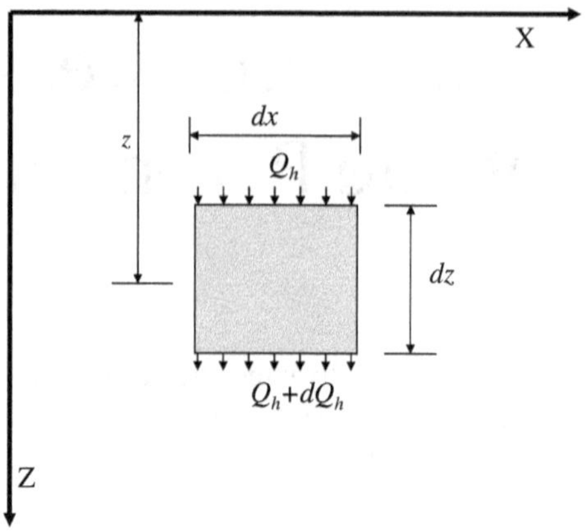

Figure 4.1 Schematic diagram showing one-dimensional steady-state heat flow across a medium

where k^{td} is the coefficient of thermal diffusivity. Putting, Equation 4.3 in Equation 4.2, one obtains

$$k^{td}\frac{d^2T}{dz^2} = \frac{dT}{dt} \tag{4.4}$$

Solution of Equation 4.4 provides the thermal profile of the material, $T(z)$ across its depth. Since the present formulation is one dimensional, at a given value of z, the $T(z)$ value would remain the same for any (x,y). Given that the typical dimension of a pavement $(L \times B)$ is considerably large with respect to its height (h), this may be a reasonable assumption. Past studies (theoretical as well as experimental) suggest that thermal profile $(T(z))$ in the concrete slab is generally nonlinear for most of time during day and night [21, 58, 59, 175, 187, 245, 337].

For a multi-layered structure, the Equation 4.4 can be presented as [316]

$$k_i^{td}\frac{d^2T_i}{dz^2} = \frac{dT_i}{dt} \tag{4.5}$$

where k_i^{td} presents the thermal diffusivity of the ith layer, and T_i represents the thermal profile of the ith layer. Equation 4.5 can be solved using appropriate boundary conditions. These are discussed in the following.

4.2.1 SURFACE BOUNDARY CONDITION

The ambient air temperature adjacent to the pavement may be assumed to be varying in a sinusoidal pattern with 24 hours as the cycle length. Accordingly, the pavement surface temperature also may be assumed to vary in a sinusoidal pattern [143, 175]. Thus, the surface boundary condition can be written as

$$T_1^{mboxtop} = T_a + T_f sin\omega_f t \tag{4.6}$$

where T_a is the mean pavement temperature, T_f is the amplitude of temperature variation, ω_f is the frequency of sinusoidal variation. Alternatively, the pavement surface temperature can be modelled [109] considering the exchange of heat between the pavement surface and surrounding through convection and radiation.

4.2.2 INTERFACE CONDITION

The temperatures at the interface should be the same. That is, the temperature at the bottom of the ith layer should be equal to the temperature at the top of the $i+1$th layer. That is, for an interface of ith and $i+1$th layers, it can be written as [175, 316]

$$T_i^{\text{bottom}} = T_{i+1}^{\text{top}} \tag{4.7}$$

Since at the interface, heat flow should be equal, one can write [109, 175, 316]

$$-k_i^{td}\rho_i C_i^h \left.\frac{dT_i}{dz}\right|_{\text{bottom}} = -k_{i+1}^{td}\rho_{i+1}C_{i+1}^h \left.\frac{dT_{i+1}}{dz}\right|_{\text{top}} \tag{4.8}$$

where ρ_i, C_i^h and k_i^{td}, represent the density, heat capacity and coefficient of thermal diffusivity of the ith layer, respectively.

4.2.3 CONDITION AT INFINITE DEPTH

The temperature at the bottom of the nth layer (which is a half-space, refer to Section 5.3 for more discussions on half-space) may be assumed as constant (say, T_∞) [175]. Thus,

$$T_n^{\text{bottom}} = T_\infty \tag{4.9}$$

The above boundary conditions can be used for obtaining the thermal profile in a layered pavement structure. One may, for example, refer to [175] for a closed formed solution for a three-layered structure.

4.3 THERMAL STRESS IN CONCRETE PAVEMENT

As mentioned earlier (refer Section 4.1), the resistance to movement (which evolves due to thermal variations) causes thermal stress. This resistance to the movement may be provided by the self weight of the concrete slab (when the slab is trying to bend upward), the underlying layer (when the slab is trying to bend downward) and friction between the slab and the underlying layer (when the slab is trying to move horizontally). If T_o is assumed as the temperature at which the pavement is stress-free, one can write

$$\varepsilon_{xx}^T = \varepsilon_{yy}^T = -\alpha\left(T(z) - T_o\right) \tag{4.10}$$

where α is the coefficient of thermal expansion. It is interesting to note that T_o may not necessarily be a uniform temperature, as assumed in Equation 4.10. In fact, experimental studies show that T_o rather assumes a non-uniform temperature distribution [193]. It may be noted that negative sign has been used to indicate this strain will

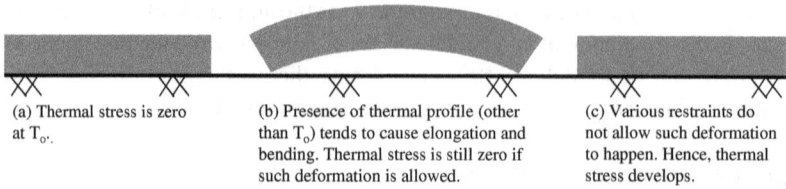

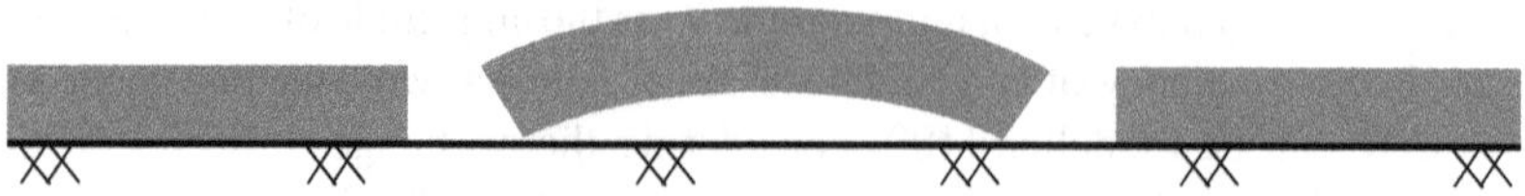

(a) Thermal stress is zero at T_o.

(b) Presence of thermal profile (other than T_o) tends to cause elongation and bending. Thermal stress is still zero if such deformation is allowed.

(c) Various restraints do not allow such deformation to happen. Hence, thermal stress develops.

Figure 4.2 A one-dimensional schematic representation of development of thermal stress in concrete slab

generate compressive stress (with an assumption that $T(z) > T_o$ for any value of z), if restrained. The concept of development of thermal stress has been schematically presented in Figure 4.2.

4.3.1 THERMAL STRESS UNDER FULL RESTRAINED CONDITION

If the strain represented as Equation 4.10 is restrained, thermal stress will originate. The thermal stress (under fully restrained condition), therefore, can be written [238] as (refer to Equation 1.34 for plane stress condition),

$$\sigma_{xx}^T = \frac{E}{1-\nu^2}\varepsilon_{xx}^T + \nu\frac{E}{1-\nu^2}\varepsilon_{yy}^T = -\frac{E\alpha(T(z)-T_o)}{1-\nu} \tag{4.11}$$

$$\sigma_{yy}^T = \frac{E}{1-\nu^2}\varepsilon_{yy}^T + \nu\frac{E}{1-\nu^2}\varepsilon_{xx}^T = -\frac{E\alpha(T(z)-T_o)}{1-\nu} \tag{4.12}$$

Thus, it can be seen that (with the current assumption that thermal profile only varies along Z direction) the thermal stress $\sigma_{xx}^T = \sigma_{yy}^T = \sigma^T$ (say). The moment due to temperature stress can be calculated[2] as follows:

$$\begin{aligned} M_{xx}^T = M_{yy}^T &= -\int_{-h/2}^{h/2} \frac{E\alpha(T(z)-T_o)}{1-\nu} z\,dz \\ &= -\frac{E\alpha}{1-\nu}\int_{-h/2}^{h/2} T(z) z\,dz = M^T \text{ (say)} \end{aligned} \tag{4.13}$$

The thermal stress (Equations 4.11 and 4.12) can be broken-up into three components as axial, bending and nonlinear [58]. This analysis becomes useful because, due to provisions of joints (where horizontal movement is allowed) the axial stress may get dissipated, and the design is governed by the bending and remaining nonlinear components of stresses. For a given total thermal stress ($\sigma_{xx}^T = \sigma_{yy}^T = \sigma^T$) in concrete pavement slab (under fully restrained condition) due to thermal profile ($T(z)$) (refer Equations 4.11 and 4.12), the following sections discuss the approach to obtain the axial (σ^{TA}), bending (σ^{TB}) and nonlinear component (σ^{TN}) of the stress [58, 132].

[2]Assuming E and α are not affected by temperature, and hence can be taken outside the integration sign.

4.3.1.1 Axial Stress Component

One can assume an equivalent axial stress component (σ^{TA}), due to equivalent axial (uniform) temperature profile as T^A. Equating the total thermal force (for unit width of the slab) due to thermal profile $T(z)$ with that of T^A, one obtains [58, 132, 337]

$$\int_{-h/2}^{h/2} \sigma^{TA} dz = \int_{-h/2}^{h/2} \sigma^T dz \tag{4.14}$$

Considering, $\sigma^{TA} = \frac{E\alpha}{(1-\nu)}(T^A - T_o)$ and $\sigma^T = \frac{E\alpha}{(1-\nu)}(T(z) - T_o)$ (refer Equations 4.11 and 4.12), one obtains

$$T^A = \frac{1}{h}\int_{-h/2}^{h/2} T(z)dz \tag{4.15}$$

Thus, the axial component of the thermal stress (σ^{TA}) is

$$\begin{aligned}\sigma^{TA} &= -\frac{E\alpha}{1-\nu}(T^A - T_o)\\ &= -\frac{E\alpha}{1-\nu}\left(\frac{1}{h}\int_{-h/2}^{h/2} T(z)dz - T_o\right)\end{aligned} \tag{4.16}$$

Equation 4.16 provides an estimate for the axial component of thermal stress under full restraint condition. If it is assumed that the frictional force provides resistance to the movement, one can write

$$\sigma^{TA} = \rho g f L \tag{4.17}$$

where ρ = density of concrete, f = coefficient of friction, L = length of concrete slab. However, it is argued that when the slab is trying to expand/contract, full frictional force may not be realized throughout the entire slab, especially near the central portion of the slab where the force developed may be lower than the maximum frictional resistance. Hence, for design purpose, the stress may be reduced by some factor [197, 290, 328, 343].

Because of the provision of the expansion joints, the axial stress may get a scope for dissipation. However, the restraint provided by the underneath layer may partial and the slab may manage to move partially (in that case, stress will be lower than full restraint case). A formulation for partial axial restraint has been dealt further in Section 4.3.2.1.

4.3.1.2 Bending Stress Component

One can assume an equivalent bending stress (σ^{TB}) due to equivalent bending (linear) temperature profile, T^B. Equating the total moment due to thermal stress (for unit width of slab) due to thermal profile $T(z)$, with that of T^B one obtains [58, 132],

$$\int_{-h/2}^{h/2} \sigma^{TB} z dz = \int_{-h/2}^{h/2} \sigma^T z dz \tag{4.18}$$

Considering, $\sigma^{TB} = -\frac{E\alpha}{(1-\nu)}(T^B - T_o)$, $\sigma^T = -\frac{E\alpha}{(1-\nu)}(T(z) - T_o)$ (refer Equation 4.11), and $\frac{T^B - T_o}{z}$ is a constant parameter (because an equivalent linear profile is being considered here), one obtains

$$T^B = T_o + \frac{12z}{h^3}\int_{-h/2}^{h/2} T(z) z dz \tag{4.19}$$

Thus, the bending component of the thermal stress (σ^{TB}) is

$$\sigma^{TB} = -\frac{E\alpha}{1-\nu}\left(T^B - T_o\right) = -\frac{12Ez\alpha}{(1-\nu)h^3}\int_{-h/2}^{h/2} T(z) z dz \tag{4.20}$$

The bending moment can be expressed as (refer to Equations 4.11, 3.27 and 3.28)

$$\begin{aligned} M_{xx}^{TB} = M_{yy}^{TB} &= -\int_{-h/2}^{h/2} \frac{E\alpha\left(T^B - T_o\right)}{1-\nu} z dz \\ &= -\frac{E\alpha}{1-\nu}\int_{-h/2}^{h/2} T^B z dz = M^{TB} \text{ (say)} \end{aligned} \tag{4.21}$$

Equation 4.20 is the expression for bending stress under full restraint condition due to the bending component of the thermal profile. The self-weight of the concrete slab provides this restraint. However, restraint provided by the self-weight may be partial, and the slab may manage to move partially (in that case stress will be lower than full restraint case). A formulation for partial bending restraint has been dealt further in Section 4.3.2.2.

4.3.1.3 Nonlinear Stress Component

Since the thermal profile $T(z)$ may have an arbitrary shape, the axial (σ^{TA}) and bending stress (σ^{TB}) added together may not necessarily make up to the total stress. Thus, the nonlinear component of stress, σ^{TN} can be obtained by subtracting the sum of σ^{TA} and σ^{TB} from total stress σ^T. Thus [58, 132, 238],

$$\begin{aligned} &\sigma^{TN} = \sigma^T - \left(\sigma^{TA} + \sigma^{TB}\right) \\ &\text{Or,} = -\frac{E\alpha}{1-\nu}\left(T(z) - \frac{1}{h}\int_{-h/2}^{h/2} T(z) dz - \frac{12z}{h^3}\int_{-h/2}^{h/2} T(z) z dz\right) \end{aligned} \tag{4.22}$$

The nonlinear component of temperature T^N can be calculated as [132]

$$\begin{aligned} \left(T^N - T_o\right) &= \left(T(z) - T_o\right) - \left(T^A - T_o\right) - \left(T^B - T_o\right) \\ T^N &= T_o + T(z) - \frac{1}{h}\int_{-h/2}^{h/2} T(z) dz - \frac{12z}{h^3}\int_{-h/2}^{h/2} T(z) z dz \end{aligned} \tag{4.23}$$

Thus, a simple formulation for calculation of T^A, T^B, T^N and corresponding σ^{TA}, σ^{TB} and σ^{TN} is presented in the above for single concrete layer under plane stress case. For further studies on two slabs with and without bonding, one can refer to [132].

4.3.1.4 Example Problem

Thermal profile of a concrete pavement slab is given as

$$T(z) = \frac{T_t + T_b}{2} - \frac{T_t - T_b}{h} z \tag{4.24}$$

where T_t and T_b are the temperatures at the top and bottom of the concrete slab, respectively. Estimate the axial (σ^{TA}) and bending (σ^{TB}) component of the thermal stress. Show that the nonlinear thermal stress (σ^{TN}) is nil in this case [238].

4.3.1.5 Solution

Incorporating $T(z) \left(= \frac{T_t+T_b}{2} - \frac{T_t-T_b}{h} z \right)$ in Equation 4.16, one obtains

$$\sigma^{TA} = -\frac{E\alpha}{1-\nu} \left(\frac{T_t+T_b}{2} - T_o \right) \tag{4.25}$$

Similarly, incorporating $T(z)$ in Equation 4.20, one obtains

$$\sigma^{TB} = \frac{E\alpha z}{h(1-\nu)} (T_t - T_b) \tag{4.26}$$

From Equation 4.11, the total stress is calculated as

$$\sigma^{T} = -\frac{E\alpha}{1-\nu} \left(\frac{T_t + T_b}{2} - \frac{T_t - T_b}{h} z - T_o \right) \tag{4.27}$$

It can be seen, that for the present case

$$\sigma^T = \sigma^{TA} + \sigma^{TB}$$

That is, $\sigma^{TN} = 0$

4.3.1.6 Discussions

From Equation 4.26, the bending stress at the top fibre of the slab is calculated as $\sigma^{TB}|_{z=-h/2} = -\frac{E\alpha(T_t-T_b)}{2(1-\nu)}$. This means that during day time when $T_t > T_b$, the thermal stress (bending) is compressive (as per the sign convention adopted) at the top. This is what is expected. The thermal profile and the stress profile (for $T_t > T_b$) are shown in Figure 4.3. During the day time, the top portion of the slab tries to expand more than the bottom portion. However, self-weight restrains it from bending, causing compressive stress to develop at the top. Using the same logic, it can be said that the stress will be tensile at the bottom portion (during day time, when $T_t > T_b$). The formulation for bending stress for linear thermal profile under fully restrained condition (infinite slab) was originally derived by Westergaard [130, 175, 318].

Figure 4.4 presents two possible situations (for $T_t > T_b$) when the slab is assumed to be resting on a rigid base (refer Figure 4.2). If the finite slab is fully restrained (for example, if the unit weight is infinite), σ^{TB} will keep on increasing with the

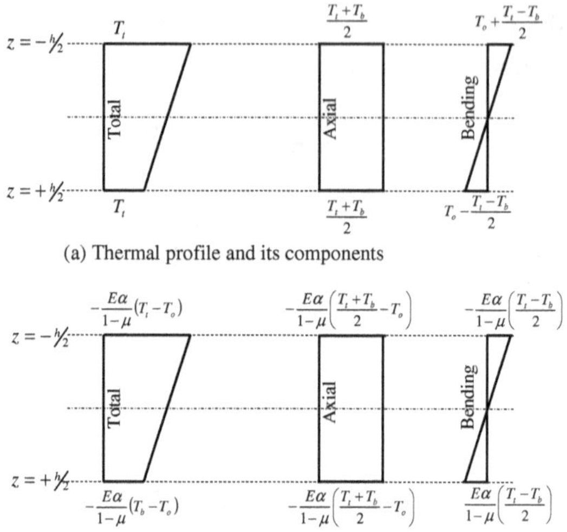

Figure 4.3 The thermal profile and thermal stress components for the example problem (when $T_t > T_b$)

increase of $T_t - T_b$ (refer Figure 4.4). If it is assumed that the self-weight slab is finite, then σ^{TB} will keep on increasing with the increase of $T_t > T_b$ up to a certain threshold value, say $(T_t - T_b)^*$, and after that σ^{TB} would assume a fixed value (refer Figure 4.4). This is because the self-weight could provide the maximum restriant that it could provide, and hence σ^{TB} would not increase further with the increase of $T_t - T_b$. A simple formulation considering the combined effect of bending due to temperature and self-weight is presented in Section 4.3.2.2.

4.3.2 THERMAL STRESS UNDER PARTIALLY RESTRAINED CONDITION

The restraints cause hindrances to the free movement of the slab due to temperature change. The resistance provided by the restraints (against thermal movements) may have finite limits; hence, the slab may become partially restrained. Figure 4.5 shows a schematic diagram indicating the shapes of the concrete slab under full and partial[3] restraint conditions.

In the following, formulations for thermal stress due to partial restraint (for axial and bending) are discussed. The solutions converge to full-restraint situation, if the movement of the slab is assumed to be zero, that is, $u = 0$ (for axial restraint case) or $w = 0$ (for bending restraint case).

4.3.2.1 Partial Axial Restraint

The axial restraint is provided by the layer underneath the concrete slab. Free body diagram of an element of length dx and width B of the slab is shown in Figure 4.6,

[3]one of the possible shapes.

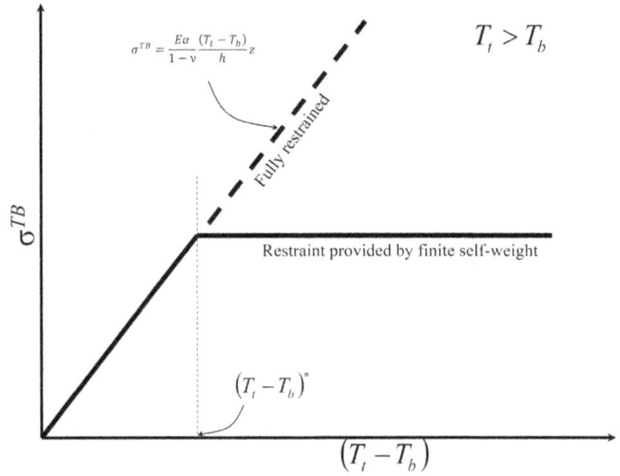

Figure 4.4 Schematic diagram showing variation σ^{TB} for $T_t > T_b$ when (i) fully restrained and (ii) partially restrained due to finite self-weight

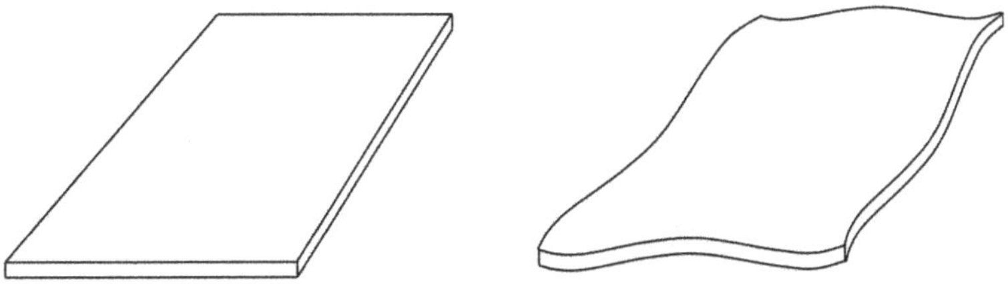

(a) Slab under fully restrained condition (b) Slab under partially restrained condition

Figure 4.5 Shape of the slab under full and partial restraint condition

where τ_{zx} is the shear stress at the interface. From force equilibrium, it can be written [52, 264, 283],

$$\tau_{zx} B dx = d\sigma^{TA} Bh \tag{4.28}$$

where dx is an elemental length, τ_{zx} = shear stress at the interface, B is the width of the concrete slab, h is the height of the concrete slab. It is considered that the pavement is supported by spring sliders (of spring constant k_{ss}) as shown in Figure 4.7. Substituting,

$$\tau_{zx} = k_{ss} u \tag{4.29}$$

in Equation 4.28, it can be written as [52, 264]

$$\frac{d\sigma^{TA}}{dx} = \frac{k_{ss} u}{h} \tag{4.30}$$

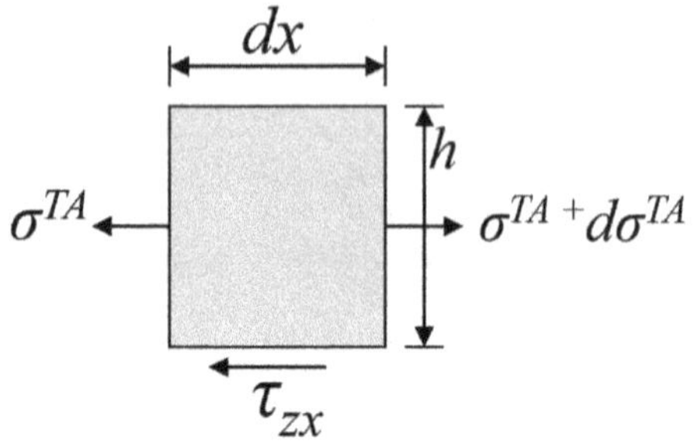

Figure 4.6 Free body diagram of an element of length dx and width B of the slab

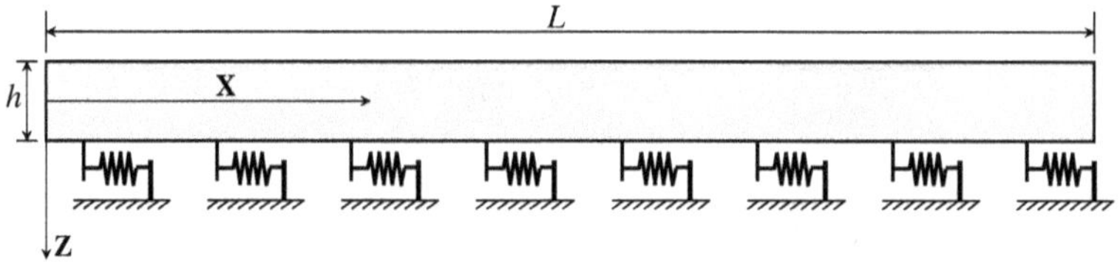

Figure 4.7 Slab resting on spring sliders

It may be noted that the k_{ss} represents similar to horizontal spring constant of the underlying layer. Interested readers may refer to [282], for example, for discussions on physical interpretation and evaluation of this parameter. Since the thermal stress is generated from the restrained strain (partial restrained strain in the present case, because some displacement (u) is being allowed), one can write [197, 283]

$$\sigma^{TA} = E\left(\frac{du}{dx} - \alpha(T^A - T_o)\right) \tag{4.31}$$

Combining Equations 4.30 and 4.31, one obtains [52, 124]

$$\frac{d^2u}{dx^2} - \frac{k_{ss}}{hE}u = 0 \tag{4.32}$$

This equation can be solved to obtain the value of u and subsequently, the value of σ^{TA} can be obtained by using Equation 4.31.

Since the length of the slab is L, it can be assumed that $\sigma_{xx}|_{x=0} = 0$, and $u|_{x=L/2} = 0$. With these boundary conditions, the solution obtained by Timm et al. [283] can be presented as Equation 4.32,

$$u = \alpha\left(T^A - T_o\right)\frac{e^{-\xi(L-x)} - e^{-\xi x}}{\xi(1+e^{-\xi L})} \tag{4.33}$$

$$\sigma_{xx} = E\alpha\left(T^A - T_o\right)\left[\frac{e^{-\xi(L-x)} + e^{-\xi x}}{1+e^{-\xi L}} - 1\right] \tag{4.34}$$

for, $0 \leq x \leq L/2$ where, $\xi = \sqrt{\frac{k_{ss}}{Eh}}$

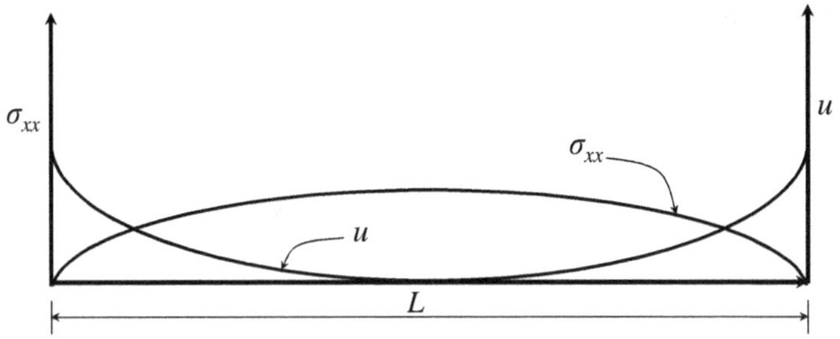

Figure 4.8 A schematic diagram representing the variation of σ_{xx} and u along the length of the slab for partial restraint condition in model shown in Figure 4.7

A schematic diagram showing variations of σ_{xx} and u for such a model are shown in Figure 4.8. In case a friction model (with a partial slippage) is used, the Equation 4.32 can be solved differently, and one can refer to [338] for a suggested approach.

4.3.2.2 Partial Bending Restraint

The self-weight of the slab (or any other external loading applied to the slab), the hypothetical springs (as foundation) and the bending due to temperature profile determine the final shape (hence the degree of restraint) and subsequently the overall stress in the pavement (refer Figure 4.5(b)). The equivalent bending temperature (T^B) has been considered in the following formulation. It may be noted that one can use $T(z)$ as well (in that case, it will not represent pure bending). Considering Equations 4.10 and 3.25, one can write [175]

$$\begin{aligned} \varepsilon_{xx} &= -z\frac{\delta^2 w}{\delta x^2} - \alpha\left(T^B - T_o\right) \\ \varepsilon_{yy} &= -z\frac{\delta^2 w}{\delta y^2} - \alpha\left(T^B - T_o\right) \\ \gamma_{xy} &= -2z\frac{\delta^2 w}{\delta x \delta y} \end{aligned} \tag{4.35}$$

Referring to Equation 3.26, one obtains [132, 175, 309]

$$\begin{aligned} \sigma_{xx} &= -\frac{Ez}{1-\nu^2}\left(\frac{\delta^2 w}{\delta x^2} + \nu\frac{\delta^2 w}{\delta y^2}\right) - \frac{E\alpha(T^B - T_o)}{1-\nu} \\ \sigma_{yy} &= -\frac{Ez}{1-\nu^2}\left(\frac{\delta^2 w}{\delta y^2} + \nu\frac{\delta^2 w}{\delta x^2}\right) - \frac{E\alpha(T^B - T_o)}{1-\nu} \\ \tau_{xy} &= \frac{E\gamma_{xy}}{2(1+\nu)} = -\frac{Ez}{1+\nu}\frac{\delta^2 w}{\delta x \delta y} \end{aligned} \tag{4.36}$$

Considering Equations 3.27, 3.28 and 3.29, one can write [175,266,318]

$$\begin{aligned} M_{xx} &= -D\left[\frac{\delta^2 w}{\delta x^2}+\nu\frac{\delta^2 w}{\delta y^2}\right]-\frac{E\alpha}{1-\nu}\int_{-h/2}^{h/2}\left(T^B-T_o\right)zdz \\ M_{yy} &= -D\left[\frac{\delta^2 w}{\delta y^2}+\nu\frac{\delta^2 w}{\delta x^2}\right]-\frac{E\alpha}{1-\nu}\int_{-h/2}^{h/2}\left(T^B-T_o\right)zdz \\ M_{xy} &= D(1-\nu)\frac{\delta^2 w}{\delta x\delta y} \end{aligned} \tag{4.37}$$

Putting Equation 4.37 in Equation 3.35, recalling Equation 4.21 and further simplification yields

$$D\nabla^4 w+\nabla^2 M^{TB}=q^\star \tag{4.38}$$

where $M^{TB} = -\frac{E\alpha}{1-\nu}\int_{-h/2}^{h/2}\left(T^B-T_o\right)zdz = -\frac{E\alpha}{1-\nu}\int_{-h/2}^{h/2}\left(T^B\right)zdz$ (that is, same as Equation 4.21). It may be recalled that it has been assumed that T^B (so also, $T(z)$) varies only along Z direction; hence, M^{TB} is only a function of z. Thus, $\nabla^2 M^{TB}=0$. Therefore, the equation takes the following form

$$D\nabla^4 w=q^\star$$

which is same as Equation 3.36. Thus, it is interesting to note that the thermal profile does not show up in the final equation; however, it participates in the solution when appropriate boundary conditions are invoked during solving the equation. That is, the moment expressions (refer Equation 4.37) have thermal profile term, and depending on the boundary condition (free, hinged or fixed), these conditions participate in the solution process of the equation. One can refer to Section 3.3.4 to see how the boundary conditions are used with reference to the calculation of load stress in plate.

The closed form solution of the above equation (for partial restraint case) is however a complex task. For 1-D formulation (beam case) one can refer [280]. This work also provides derivations for the situation when the slab is curled up for the case when $T_t < T_b$ (that is, with loss of contact) and the springs become ineffective for that region [280].

For slabs with finite dimensions, it was suggested that the deflected shape (w) can be represented by superposition of deflections of the slab when it is assumed to be finite in one direction and infinite along other direction [130, 280, 318]. And subsequently, the proposed solutions for maximum stresses at the interior (at $B/2$, $L/2$) for linear thermal profile are

$$\sigma_{xx}=\frac{E\alpha(T_t-T_b)}{2(1-\nu^2)}(C_x+\nu C_y) \tag{4.39}$$

$$\sigma_{yy}=\frac{E\alpha(T_t-T_b)}{2(1-\nu^2)}(C_y+\nu C_x) \tag{4.40}$$

where C_x and C_y are two coefficients related to the shape of the slab [36, 130, 280, 318].

4.4 CLOSURE

Simple formulations for computation of thermal stress (for partial and full restraint) for concrete pavement has been presented in this chapter. Formulas or charts for estimation of thermal stress are available in various documents and guidelines [1, 131, 214, 290].

5 Load Stress in Asphalt Pavement

5.1 INTRODUCTION

It is in general assumed that asphalt pavements do not have any bending moment carrying capacity unlike concrete pavement. Thus, the equilibrium equations which are developed to estimate stresses in asphalt pavements should not contain any bending moment term. Hence, bituminous pavements are modelled as layered continuum media. This has been detailed in this chapter.

The simplest model one can think of is a single layer continuum media. For example, a large ground infinitely extended along the Z and the R directions (in cylindrical coordinate system) made up of homogeneous, isotropic, linear elastic soil can be idealized as a continuum media occupying half of the entire space, hence known as half-space.

In this chapter, first, the approach to analyse a single layer continuum (that is, a half-space) is presented. Subsequently, the approach is extended for a muti-layered case (with smooth or rough interface), representing an idealized asphalt pavement structure.

5.2 GENERAL FORMULATION

To formulate a problem (so as to obtain its mechanical response), one needs to have (i) strain displacement relationship (hence strain compatibility conditions), (ii) stress and strain relationship (that is constitutive relationship), (iii) equilibrium condition and the (iv) geometry (that is the boundary conditions) of the problem (refer Section 1.2 for the background information). In the following an expression is developed as a combined form of the first three considerations. Finally, the boundary conditions will be used to analyze (as a typical boundary value problem) a half-space, and subsequently, it will be extended to solve a multi-layered structure.

Let a plane strain condition (X–Z plane) in Cartesian coordinate be assumed. The strains (ε_{xx}, ε_{zz} and γ_{xz}) can be expressed as Equation 1.35. Putting the above in relevant strain compatibility equation (refer to the third equation of the set of Equations 1.20), one obtains

$$\frac{\partial^2}{\partial x \partial z}\left(\frac{2(1+\nu)}{E}\tau_{zx}\right) = \frac{\partial^2}{\partial x^2}\left(\frac{1-\nu^2}{E}\sigma_{zz} - \frac{\nu(1+\nu)}{E}\sigma_{xx}\right) + \frac{\partial^2}{\partial z^2}\left(\frac{1-\nu^2}{E}\sigma_{xx} - \frac{\nu(1+\nu)}{E}\sigma_{zz}\right) \quad (5.1)$$

DOI: 10.1201/9781003190769-5

Or,

$$2(1+\nu)\frac{\partial^2\tau_{zx}}{\partial x\partial z} = (1-\nu^2)\frac{\partial^2\sigma_{zz}}{\partial x^2} - \nu(1+\nu)\frac{\partial^2\sigma_{xx}}{\partial x^2} + (1-\nu^2)\frac{\partial^2\sigma_{xx}}{\partial z^2} - \nu(1+\nu)\frac{\partial^2\sigma_{zz}}{\partial z^2} \tag{5.2}$$

Now, the equilibrium equation (refer to Equation 1.39) for the present case (considering a plane strain along the X-Z plane) can be written as

$$\frac{\partial\sigma_{xx}}{\partial x} + \frac{\partial\tau_{zx}}{\partial z} + BF_x = 0 \tag{5.3}$$

$$\frac{\partial\tau_{zx}}{\partial x} + \frac{\partial\sigma_{zz}}{\partial z} + BF_z = 0 \tag{5.4}$$

Differentiating Equation 5.3 with respect to x and differentiating Equation 5.4, with respect to z, and adding one obtains

$$2\frac{\partial^2\tau_{zx}}{\partial z\partial x} = -\left(\frac{\partial^2\sigma_{xx}}{\partial x^2} + \frac{\partial^2\sigma_{zz}}{\partial z^2}\right) - \left(\frac{\partial BF_x}{\partial x} + \frac{\partial BF_z}{\partial z}\right) \tag{5.5}$$

Substituting $2\frac{\partial^2\tau_{zx}}{\partial z\partial x}$ from Equation 5.5 to Equation 5.2 and simplifying further, one obtains

$$\nabla^2(\sigma_{xx} + \sigma_{zz}) = -\frac{1}{1-\nu}\left(\frac{\partial BF_x}{\partial x} + \frac{\partial BF_z}{\partial z}\right) \tag{5.6}$$

where $\nabla^2 = \left(\frac{\partial^2}{\partial x^2} + \frac{\partial^2}{\partial z^2}\right)$

Proceeding in similar manner for plane stress condition (that is, using Equations 1.34 instead of Equations 1.35 and considering X–Z plane), one obtains

$$\nabla^2(\sigma_{xx} + \sigma_{zz}) = -(1+\nu)\left(\frac{\partial BF_x}{\partial x} + \frac{\partial BF_z}{\partial z}\right) \tag{5.7}$$

Equation 5.7 is also known as stress compatibility equation. Now, if the body forces are assumed to be zero (if weight is the only body force considered, then $BF_x = 0$, further, if the medium is also considered as weightless, then $BF_z = 0$), then the equations reduce to the form

$$\nabla^2(\sigma_{xx} + \sigma_{zz}) = 0 \tag{5.8}$$

Assuming, ϕ is a function of x and z which can be expressed in the following form (so that it follows equilibrium equation in two dimensions, without body-forces) [181],

$$\begin{aligned} \sigma_{xx} &= \frac{\partial^2\phi}{\partial z^2} \\ \sigma_{zz} &= \frac{\partial^2\phi}{\partial x^2} \\ \tau_{xz} &= -\frac{\partial^2\phi}{\partial x\partial z} \end{aligned} \tag{5.9}$$

Then (by putting Equation 5.9, in Equation 5.8)

$$\nabla^4 \phi = 0 \tag{5.10}$$

The function ϕ is known as Airy's stress function. For a cylindrical coordinate system and axi-symmetrical case (that is, $\frac{\partial}{\partial \theta} = 0$ in the present case, refer to Section 1.2.3 for further discussions), the stresses in terms of differentiation of ϕ function can be expressed as [181, 230, 288]

$$\sigma_{zz} = \frac{\partial}{\partial z}\left((2-\nu)\nabla^2\phi - \frac{\partial^2\phi}{\partial z^2}\right) \tag{5.11}$$

$$\sigma_{rr} = \frac{\partial}{\partial z}\left(\nu\nabla^2\phi - \frac{\partial^2\phi}{\partial r^2}\right) \tag{5.12}$$

$$\sigma_{\theta\theta} = \frac{\partial}{\partial z}\left(\nu\nabla^2\phi - \frac{1}{r}\frac{\partial\phi}{\partial r}\right) \tag{5.13}$$

$$\tau_{rz} = \frac{\partial}{\partial r}\left((1-\nu)\nabla^2\phi - \frac{\partial^2\phi}{\partial z^2}\right) \tag{5.14}$$

Further from Equations 1.21 and 1.30, and considering axi-symmetric case,

$$\begin{aligned} w &= \frac{1+\nu}{E}\left[(1-2\nu)\nabla^2\phi + \frac{\partial^2\phi}{\partial r^2} + \frac{1}{r}\frac{\partial\phi}{\partial r}\right] \\ u &= -\frac{1+\nu}{E}\frac{\partial^2\phi}{\partial r\partial z} \end{aligned} \tag{5.15}$$

Proceeding in the similar manner, one obtains

$$\nabla^4 \phi = 0 \tag{5.16}$$

where $\nabla^2 = \frac{\partial^2}{\partial r^2} + \frac{1}{r}\frac{\partial}{\partial r} + \frac{\partial^2}{\partial z^2}$.

The equation (Equation 5.10 or Equation 5.16) represents a combined form of strain compatibility, constitutive relationship (both for plane strain or plane stress or axi-symmetric case) and the equilibrium condition. Interested readers can refer to for example, Poulos and Davis [230] for the equations that arise for a 3-D linear elastic case.

The following steps are involved to obtain the stresses (and subsequently the strains and displacements) of such a continuum media.

- Obtain a suitable ϕ function which satisfies Equation 5.10 or Equation 5.16 depending on the choice of coordinate system.
- Obtain the constants of the ϕ function from the boundary conditions of a given problem.
- Obtain the stresses from Equation 5.9 or from Equations 5.11 to 5.14 depending on the choice of coordinate system.

The above approach can be used to solve stresses in continuum for various loading and geometries. This has been discussed further in the following.

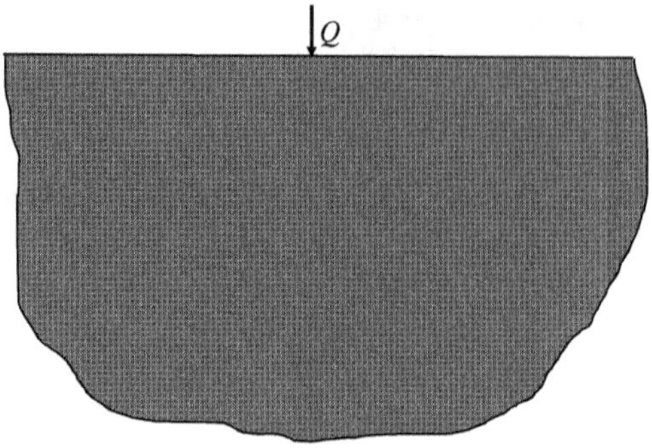

Figure 5.1 A concentrated load acting vertically on a half-space

5.3 SOLUTION FOR ELASTIC HALF-SPACE

Figure 5.1 shows a concentrated load (point load) of magnitude of Q acting vertically on an elastic half-space, and the objective is to find out the stress–strain displacement of any point (r, z) within the half-space. This problem is commonly known as Boussinesq's problem, and its solution is well known and widely used in the literature. One may refer to [146] for a discussion on the history and background of the Boussinesq's problem. The boundary conditions for this problem can be identified as follows:

1. At infinity all stresses should vanish. That is, as $z \to \infty$, $\sigma_{zz} = \sigma_{rr} = \sigma_{\theta\theta} = \tau_{rz} = 0$
2. Shear stress at the surface should be zero. That is, $\tau_{rz}|_{z=0} = 0$.
3. The σ_{zz} at the surface should be zero except at the point of application of load. That is, $\sigma_{zz}|_{z=0} = 0$, except at the point of load application.
4. The sum of total force at any given horizontal plane within the half-space should be equal to Q. That is, $\int_A \sigma_{zz} dA = Q$, where A indicates area of the infinite horizontal plane.

To solve such problem, first a suitable (biharmonic) ϕ function needs to be selected. The discussions in the following are based on the solution presented in [144]. The following ϕ function is proposed [144] for the present case. It may be verified that the condition given by Equation 5.16 holds for this ϕ function.

$$\phi = C_1 z ln_e r + C_2(r^2+z^2)^{1/2} + C_3 z ln_e \frac{(r^2+z^2)^{1/2} - z}{(r^2+z^2)^{1/2} + z} \tag{5.17}$$

From the above-mentioned conditions, the constants are obtained as [144],

$$C_3 = -\frac{1-2\nu}{4\nu} C_2 C_2 = -\frac{\nu}{\pi} Q C_1 = -\frac{1-2\nu}{2\pi} Q$$

After evaluating the constants of the ϕ function, it can be used in Equations 5.11 to 5.14 to obtain the stresses. The widely used expressions for stresses (due to vertical concentrated load at the surface of an elastic half-space) can be presented as [70, 114, 144, 169],

$$\sigma_{zz} = \frac{3Q}{2\pi}\frac{z^3}{R^5} \tag{5.18}$$

$$\tau_{rz} = \frac{3Q}{2\pi}\frac{z^2 r}{R^5} \tag{5.19}$$

$$\sigma_{rr} = \frac{Q}{2\pi}\left(\frac{3zr^2}{R^5} - \frac{1-2\nu}{R(R+z)}\right) \tag{5.20}$$

$$\sigma_{\theta\theta} = \frac{Q}{2\pi}(1-2\nu)\left(\frac{1}{R(R+z)} - \frac{z}{R^3}\right) \tag{5.21}$$

where $R = (r^2+z^2)^{1/2}$.

A brief solution to the Boussinesq's problem has been presented in the above. One can reach the same solution with any other ϕ function, provided it satisfies Equation 5.10 or Equation 5.16 (depending on the choice of coordinates) and the boundary conditions yield meaningful solutions to the unknown constants. One can, for example, refer to [258, 259, 325] for more information on alternative approaches.

Further, the vertical displacement can be derived in the following manner. The vertical strain (ε_z) in cylindrical coordinate system is given as (refer Equations 1.21 and 1.30)

$$\varepsilon_{zz} = \frac{1}{E}\left(\sigma_z - \nu(\sigma_r + \sigma_\theta)\right) = \frac{\partial w}{\partial z} \tag{5.22}$$

Thus, displacement at any depth z (w) is given as

$$w = \frac{1}{E}\int_z^{\infty}\left(\sigma_{zz} - \nu(\sigma_{rr} + \sigma_{\theta\theta})\right)dz \tag{5.23}$$

Putting the expressions of σ_{zz}, σ_{rr} and $\sigma_{\theta\theta}$ from Equations 5.18, 5.20 and 5.21 (respectively) in Equation 5.23 and performing integration, one obtains the displacement due to concentrated load Q,

$$w = \frac{Q(1+\nu)}{2\pi E}\left[\frac{z^2}{(r^2+z^2)^{3/2}} + \frac{2(1-\nu)}{(r^2+z^2)^{1/2}}\right] \tag{5.24}$$

Thus, the deflection due to a concentrated load Q at surface at a distance r away measured from the load application point can be obtained, by putting $z = 0$ in Equation 5.24, as

$$w|_{z=0} = \frac{(1-\nu^2)Q}{\pi E r} \tag{5.25}$$

In line with the problem of concentrated load on half-space, there were similar such developments proposed by various researchers. For example, consider a line

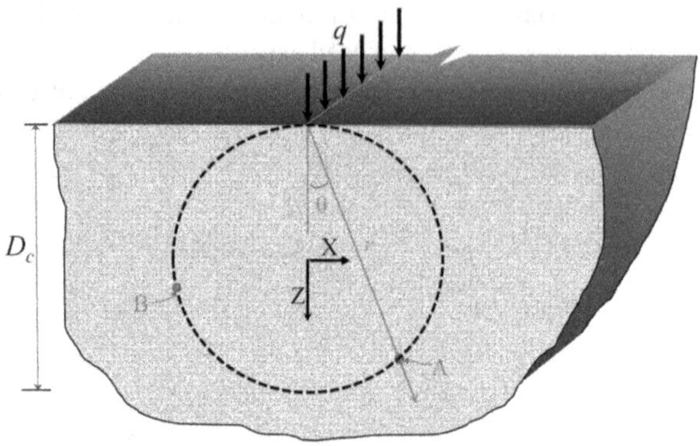

Figure 5.2 Line loading on an elastic half-space

load (q per unit length) acting vertically on surface of the half-space (Figure 5.2). This is known as Flemant's problem.

The σ_{rr} for such a geometry can be calculated at a Point A (as (r, θ), in polar coordinate system) as $-\frac{2q}{\pi}\frac{\cos\theta}{r}$. If one draws an imaginary circle of diameter D_c (as shown in Figure 5.2) which is tangent to the half-space surface and passes through the Point A, then from the property of circle, $r = D_c \cos\theta$. Thus, the final expressions for stresses are

$$\begin{aligned} \sigma_{rr} &= -\frac{2q}{\pi D_c} \\ \sigma_{\theta\theta} &= 0 \\ \tau_{r\theta} &= 0 \end{aligned} \tag{5.26}$$

In Cartesian coordinate system, the stresses are

$$\begin{aligned} \sigma_{xx} &= -\frac{2q}{\pi r}\cos\theta\sin^2\theta \\ \sigma_{zz} &= \frac{2q}{\pi r}\cos^3\theta \\ \tau_{xz} &= \frac{2q}{\pi r}\cos^2\theta\sin\theta \end{aligned} \tag{5.27}$$

As can be seen, the state of stress (expressed in polar coordinate) will be the same on another point B lying on the circle – in fact it will be same on any point on the circle.

Other variants of half-space problems could be a concentrated load acting inside an infinite space (by Kelvin) or inside a half-space (by Mindlin), concentrated load acting horizontally on the surface of a half-space (by Cerruti), etc [71, 114, 142, 144]. One can refer to, for example, [146] for an exhaustive review (and the historical context) on the contributions made by various researchers in this field and can refer to,

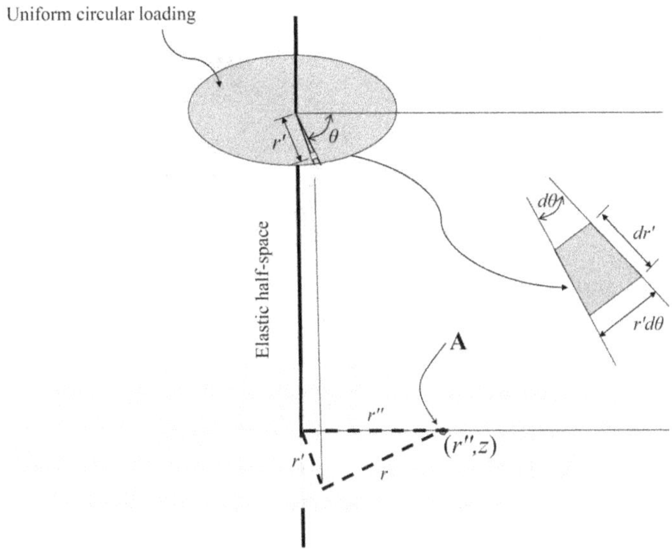

Figure 5.3 Circular loading on an elastic half-space

for example, [105] for the issues related to heterogeneous, anisotropic and incompressible continuum.

Boussinesq's approach can be extended further to find out response due to, line loading, uniform strip loading, uniform circular loading, uniform elliptical loading, triangular strip loading, Hertz loading, eccentric loading, anisotropy and heterogeneous condition, etc. Some relevant but simple example problems are discussed in the following. For further study, interested readers can refer to, for example, [45, 70, 71, 114, 142, 144, 230, 258, 288], etc.

Example problem

Circular uniform loading on elastic half-space is shown Figure 5.3. It may be assumed that a total load of Q is distributed uniformly over a circular area of radius a. Obtain an expression for σ_{zz} at a point inside the half-space (that is, at a point (r'',z) shown as A in Figure 5.3).

Solution

The stresses can be obtained assuming linear superposition is valid. That is, contributions by all infinitesimally small concentrated loads located at (r',θ) can be summed (by integration) to obtain the overall effect. The magnitude of concentrated load over an elemental area $r'\,d\theta\,dr'$ is given as $q\,r'\,d\theta\,dr'$ where $q = q_{av} = \frac{Q}{\pi a^2}$. Thus, using

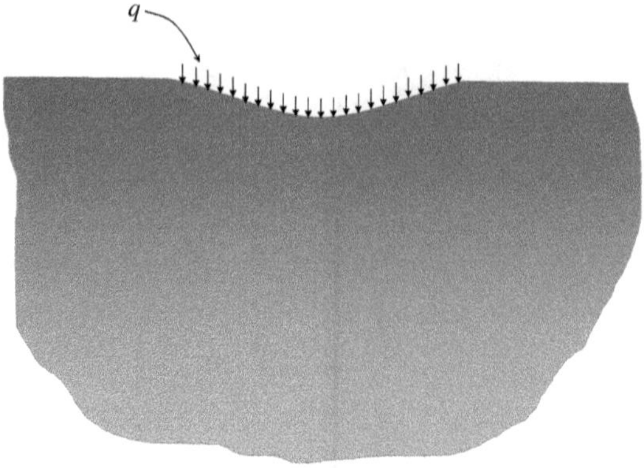

Figure 5.4 Displacement of an elastic half-space due to flexible uniformly loaded circular plate

Equation 5.18, the vertical stress (σ_{zz}), in this case, can be written as

$$\sigma_{zz} = \int_0^a \int_0^{2\pi} \frac{3(qr'd\theta dr')}{2\pi} z^3 (r^2 + z^2)^{-\frac{5}{2}} \tag{5.28}$$

where r is the distance between the elemental area to the point A. The r in the above equation can be substituted with the following relationship (from the property of triangle),

$$r = (r'^2 + r''^2 - r'r''2\cos\theta)^{1/2} \tag{5.29}$$

and integration can be performed to obtain the value of σ_{zz}. It may be noted that σ_{zz} always being along the same direction, its value due to circular loading can be directly obtained from the superposition as suggested in Equation 5.28; however, for estimation of σ_{rr} or, $\sigma_{\theta\theta}$, appropriate stress transformation would be required.

Example problem

Figure 5.4 shows a flexible uniformly loaded circular plate. Estimate the maximum surface deflection.

Solution

By flexible plate, it is understood that the plate does not have any flexural rigidity of its own, and therefore, it takes the same shape of elastic medium after deformation.

Taking an elemental area of $r'\,d\theta\,dr'$ on the loaded region, and assuming superposition is valid, one can calculate deflection w at a point (r'', 0) (using Equation 5.25) as

$$w|_{z=0} = \int_0^a \int_0^{2\pi} \frac{(1-\nu^2)(qr'd\theta\,dr')}{\pi E r} \tag{5.30}$$

where $q = q_{av} = \frac{Q}{\pi a^2}$.

Referring to Figure 5.3, where the distances r, r' and r'' have been identified, one can see that although the point under consideration is located at $(r'', 0)$, the distance between the load application point (which is the elemental area in the present case) to the point needs to be considered, and hence should be taken as r.

Since central deflection is being calculated, one notes that $r = r'$ (and $r'' = 0$) in this case. Thus,

$$w\big|_{z=0,r''=0} = \frac{2(1-\nu^2)qa}{E} \tag{5.31}$$

Assuming, $\nu = 0.5$, as a special case, one can obtain

$$w_{z=0,r''=0} = \frac{1.5q_{av}a}{E} \tag{5.32}$$

Equation 5.32 is popularly used for the estimation of elastic modulus of soil (say, subgrade or embankment structures made up of homogeneous material with sufficient thickness or height so that it can be assumed as elastic half-space) by plate load test, while a flexible plate is used.

For any point other than centre point, $r \neq r'$, and one can use the relationship given by Equation 5.29. Alternatively, one can suitably choose a different coordinate point and an elemental area to reach a solution. One can refer to, for example [71], for such an elegant solution. It can be shown that for $\nu = 0.5$, the deflection at any point on the periphery of the circular plate will be $\frac{0.95q_{av}a}{E}$, which is lower than deflection at centre.

Example problem

Figure 5.5 shows a rigid circular frictionless plate. Estimate the maximum surface deflection.

Solution

By rigid plate, it is understood that the plate has high flexural rigidity, and therefore, it does not deform itself when load is applied[1]. The pressure distribution may no longer be uniform, in this case. Two conditions need to be satisfied, as follows:

- The total force at the bottom of the plate (due to the non-uniform pressure distribution) should be equal to force Q externally applied. That is,

$$\int_A q dA = Q \tag{5.33}$$

  where A is the area of the plate, and dA is the elemental area.
- The deflection at each point of the circular plate should be the same (since it is a rigid plate).

[1]Flexible and rigid plate are obviously idealized situations – in engineering sense these are achieved by choosing suitable values of height (h) to radius (a) ratio of the plate.

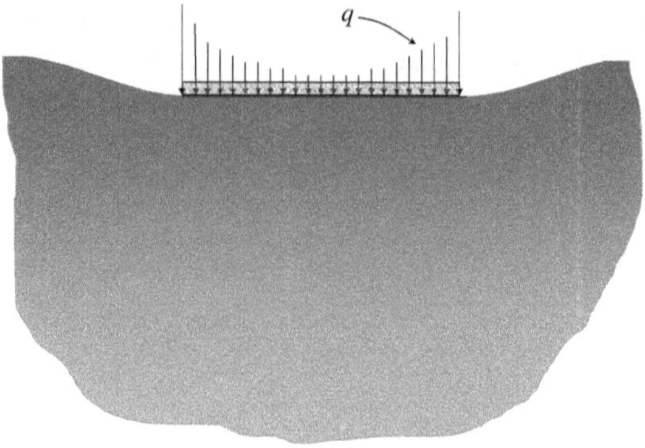

Figure 5.5 Displacement of an elastic half-space due to rigid circular plate

For the flexible plate case (that is, last example), it is found that the deflection at the centre of the plate is maximum, and at its edge, it is minimum. For the present case of rigid plate, the deflection, within the plate region, should be the same everywhere. This may possibly be achieved if the pressure is higher along the edges than at the centre [71]. Thus, the pressure distribution can be assumed to take a shape as shown in Figure 5.5. It appears that the following pressure distribution satisfies the above two conditions [71, 144].

$$q = \sigma_{zz}|_{z=0} = \frac{Q}{2\pi a(a^2 - r^2)^{1/2}} \text{ for } 0 \leq r \leq a \tag{5.34}$$

It is interesting to note that the proposed pressure distribution shows that the pressure theoretically will be infinite along the edge. One can refer to [71] for further discussions on this.

Thus, the deflection at the centre of the plate can be obtained as (derived from Equation 5.24 by putting $z = 0$, taking an elemental area of $r' d\theta\, dr'$, substituting the value of q from Equation 5.34, and recognizing $r = r'$, for the present case),

$$\begin{aligned} w|_{z=0,r''=0} &= \int_0^a \int_0^{2\pi} \frac{(1-\nu^2)(\frac{Q}{2\pi a(a^2-r^2)^{1/2}} r' d\theta\, dr')}{\pi E r} \\ &= \frac{(1-\nu^2) q_{av} a \pi}{2E} \end{aligned} \tag{5.35}$$

Assuming, $\nu = 0.5$, as a special case, one can obtain

$$w_{z=0,r''=0} = \frac{1.18 q_{av} a}{E} \tag{5.36}$$

Equation 5.36 is popularly used for the estimation of elastic modulus of soil by plate load test, while a rigid plate is used. Further calculations can show that for the

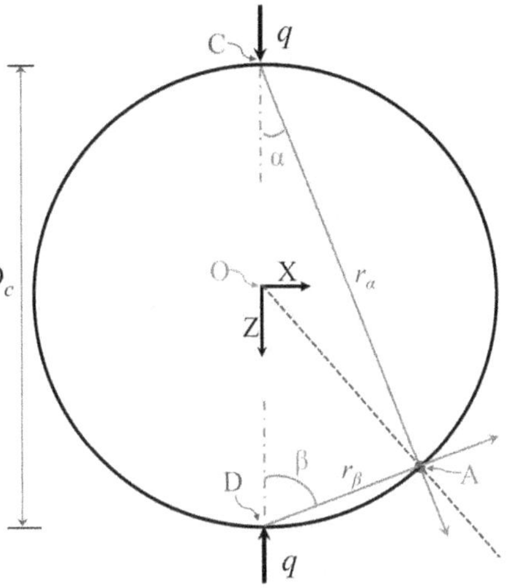

Figure 5.6 Loading applied diametrically in an IDT set-up

assumed pressure distribution (that is, Equation 5.34), the deflection at every other point within the rigid circular disk is same and can be given by Equation 5.36 for $\nu = 0.5$.

Example problem

A force q is applied per unit length diametrically to a cylindrical sample of diameter d, in an indirect test (IDT) set-up. Find out σ_{xx} along $x = 0$ within the sample.

Solution

Figure 5.6 shows a front view of the sample of unit depth. A load q is applied at Points C and D.

This problem therefore can be construed as superposition of two Flamant's problems and another loading geometry to nullify the effect of stresses on the periphery of the surface of the cylinder due to Flamant loadings [249]. This is explained in the following.

In one Flamant case a load of q is applied at Point C and the lower portion is a half-space, in another Flamant case a load of q is applied at Point D and the upper portion is a half-space.

Considering Point A, from Equation 5.26 one can write $\sigma_{rr} = -\frac{2q}{\pi D_c}$ due to load applied at Point C and the positive direction is along CA. Similarly, $\sigma_{rr} = -\frac{2q}{\pi D_c}$ at Point A due to load applied at Point D and the positive direction is along DA. The resultant traction at Point A is also found to be $-\frac{2q}{\pi D_c}$ and along the positive direction as OA. This represents a hydrostatic state of stress [249].

Thus, a stress of $\frac{2q}{\pi D_c}$ directed towards the centre along the periphery would nullify the effect of two Flamant loading, given that the stress at free surface of the cylinder should be zero[2].

Thus, from Equation 5.27 one can write [249]

$$\sigma_{xx} = -\frac{2q}{\pi r_\alpha}\cos\alpha\sin^2\alpha - \frac{2q}{\pi r_\beta}\cos\beta\sin^2\beta + \frac{2q}{\pi D_c} \quad (5.37)$$

At $x = 0$, the value of $\sigma_{xx} = \frac{2q}{\pi D_c}$, which is constant and tensile in nature. Interested readers can, for example, refer to [46] for further studies on stress distribution in an IDT set-up.

5.4 MULTI-LAYERED STRUCTURE

An idealized multi-layered asphalt pavement structure is shown in Figure 5.7. A total load of Q is assumed to act uniformly on a circular area of radius a (that is, $q = \frac{Q}{\pi a^2}$ and this is equal to the tyre-pavement contact pressure). The layers are identified as first layer, second layer, ith layer and so on, starting from top to bottom. The last layer (typically a subgrade) is the nth layer. The assumptions of a the multi-layer asphalt pavement structure can be mentioned as the following:

- The structure is constituted with n number of layers.
- Each layer is assumed to be made up of homogeneous, isotropic and linearly elastic material. Thus, elastic modulus (E_i) and Poisson's ratio (ν_i) characterize each layer.
- Each layer (except the subgrade) has finite uniform thickness (h_i), and the subgrade is assumed as semi-infinite (that is, a half-space).
- The pavement structure is assumed to be weight-less (that is, $BF_z = 0$). That is, compared to the contact pressure (q) applied, the stresses created due to self-weight of the pavement is negligible.
- The structure is also assumed to be free from any other kind of existing stresses.

5.4.1 FORMULATION

The formulation presented in Section 5.2 can be utilized to analyse a multi-layered structure. The solution approach is the same as summarized towards the end of Section 5.2; however, a transformation needs to be applied so as to match the loading geometry.

The approach for elastic analysis of multi-layered structure was originally developed by Burmister [40–42]. Various researchers have further extended the formulation (general enough to handle n layers), developed algorithms and performed numerical studies [6, 226, 257, 310, 311]. The analysis approach is presented in the following.

[2] Up till this part, for convenience, the calculation is done in polar coordinate system.

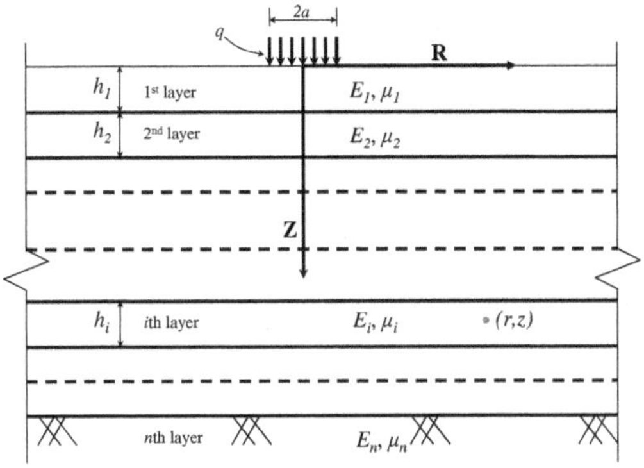

Figure 5.7 A multi-layered asphalt pavement structure

The following ϕ function is used for this analysis (and it satisfies Equation 5.16),

$$\begin{aligned}\phi^i(m) &= \left[A^i(m)e^{mz} - B^i(m)e^{-mz}\right. \\ &\quad \left. + C^i(m)ze^{mz} - D^i(m)ze^{-mz}\right] J_o(mr) \end{aligned} \tag{5.38}$$

where J_o is the Bessel function of zeroth order, $A^i(m)$, $B^i(m)$, $C^i(m)$ and $D^i(m)$ are coefficients for the ith layer, r is the radial distance and z is the depth, and m is any number. The parameter m will be useful later for the transformation purpose[3]. Thus, the form of the ϕ function remains the same, but its coefficients $A^i(m)$, $B^i(m)$, $C^i(m)$ and $D^i(m)$ change with i and m.

Using this ϕ function (that is, putting Equation 5.38 into Equation 5.11 to 5.14), the stresses (in the ith layer) and subsequently the displacement values (refer Equation 5.15), for axi-symmetric case, are calculated as follows [133,152,172,310,311]:

$$\begin{aligned}\hat{\sigma}^i_{zz} &= -mJ_o(mr)\left[A^i(m)m^2e^{mz} + B^i(m)m^2e^{-mz}\right. \\ &\quad \left. - C^i(m)m(1-2\nu_i - mz)e^{mz} + D^i(m)m(1-2\nu_i + mz)e^{-mz}\right] \\ \hat{\sigma}^i_{rr} &= mJ_o(mr)\left[A^i(m)m^2e^{mz} + B^i(m)m^2e^{-mz} + C^i(m)m(1+2\nu_i + mz)e^{mz}\right. \\ &\quad \left. - D^i(m)m(1+2\nu_i - mz)e^{-mz}\right] - \frac{J_1(mr)}{r}\left[A^i(m)m^2e^{mz}\right. \\ &\quad \left. + B^i(m)m^2e^{-mz} + C^i(m)m(1+mz)e^{mz} - D^i(m)m(1-mz)e^{-mz}\right] \\ \hat{\sigma}^i_{\theta\theta} &= \frac{J_1(mr)}{r}\left[A^i(m)m^2e^{mz} + B^i(m)m^2e^{-mz}\right. \\ &\quad \left. + C^i(m)m(1+mz)e^{mz} - D^i(m)m(1-mz)e^{-mz}\right] \\ &\quad + 2\nu_i mJ_o(mr)\left[C^i(m)me^{mz} - D^i(m)me^{-mz}\right]\end{aligned}$$

[3]Transformation is also done for the analysis of concrete pavement, and m and n are the two parameters used there. Refer to Section 3.4.

$$\begin{aligned}
\hat{\tau}^i_{rz} &= mJ_1(mr)\left[A^i(m)m^2e^{mz} - B^i(m)m^2e^{-mz}\right.\\
&\left.+C^i(m)m(2\nu_i+mz)e^{mz} + D^i(m)m(2\nu_i-mz)e^{-mz}\right]\\
\hat{w}^i &= -\frac{1+\nu_i}{E_i}mJ_o(mr)\left[A^i(m)me^{mz} - B^i(m)me^{-mz}\right.\\
&\left.-C^i(m)(2-4\nu_i-mz)e^{mz} + D^i(m)(2-4\nu_i+mz)e^{-mz}\right]\\
\hat{u}^i &= \frac{1+\nu_i}{E_i}J_1(mr)m\left[A^i(m)me^{mz} + B^i(m)me^{-mz}\right.\\
&\left.+C^i(m)(1+mz)e^{mz} - D^i(m)(1-mz)e^{-mz}\right]
\end{aligned} \tag{5.39}$$

From Equation 5.39, it can be seen that, $\hat{\sigma}^{1,t}_{zz} = -mJ_o(mr)$, where $\hat{\sigma}^{1,t}_{zz}$ represents σ_{zz} at the top of the first layer[4]. That means, the $\hat{\sigma}^i_{zz}$, $\hat{\sigma}^i_{rr}$, $\hat{\sigma}^i_{\theta\theta}$, $\hat{\tau}^i_{rz}$, $\hat{w}^i$ and $\hat{u}^i$ values calculated from Equations 5.39 are the solutions due to $mJ_o(mr)$ loading at the surface. That is why the expressions for stresses and displacements are written with a hat (^) symbol. However, one is rather interested for a solution for the case represented in Figure 5.7, that is,

$$\begin{aligned}
\sigma^{1,t}_{zz} &= -q \quad \text{for } 0 \le r \le a\\
&= 0 \quad \text{otherwise}\\
\tau^{1,t}_{rz} &= 0
\end{aligned} \tag{5.40}$$

Therefore, further transformation is needed to derive the stress and the displacement values due to loading as per Equation 5.40 (and not as per $mJ_o(mr)$). This is accomplished by invoking Henkel transform. The Hankel transform for a pair of function ($f(r)$ and $f(m)$) can be given as follows:

$$f(m) = \int_0^\infty rJ_o(mr)f(r)dr \tag{5.41}$$

$$f(r) = \int_0^\infty mJ_o(mr)f(m)dm \tag{5.42}$$

For the present case, $f(r)$ can be assumed same as $\sigma^{1,t}_{zz}$ mentioned in Equation 5.40. Thus, using Equation 5.41

$$\begin{aligned}
f(m) &= -\int_0^a qrJ_o(mr)dr + \int_a^\infty 0dr\\
&= -\frac{qa}{m}J_1(ma)
\end{aligned} \tag{5.43}$$

This expression of $f(m)$ can be put back to Equation 5.42 to re-derive an expression for $f(r)$. That is,

$$\begin{aligned}
f(r) &= \int_0^\infty mJ_o(mr)\left(-\frac{qa}{m}J_1(ma)\right)dm\\
&= qa\int_0^\infty (-mJ_o(mr))\frac{J_1(ma)}{m}dm
\end{aligned} \tag{5.44}$$

[4] superscript t indicates top and b indicates bottom

It may be recalled that the loading as $mJ_o(mr)$ is the expression obtained for $\hat{\sigma}_{zz}$ at the surface, whereas $f(r)$ presents the actual stress due to the loading shown in Figure 5.7. Thus, by taking help of the Equation 5.44, the expressions for stresses can be written as

$$\begin{aligned}
\sigma_z^i &= qa\int_0^\infty \hat{\sigma}_z^i \frac{J_1(ma)}{m}dm \\
\sigma_r^i &= qa\int_0^\infty \hat{\sigma}_r^i \frac{J_1(ma)}{m}dm \\
\sigma_\theta^i &= qa\int_0^\infty \hat{\sigma}_\theta^i \frac{J_1(ma)}{m}dm \\
\tau_{rz}^i &= qa\int_0^\infty \hat{\tau}_{rz}^i \frac{J_1(ma)}{m}dm
\end{aligned} \tag{5.45}$$

Thus, the final expressions for stresses or displacements due to loading as per Figure 5.7. Thus [133, 152, 172, 310, 311],

$$\begin{aligned}
\sigma_{zz}^i &= qa\int_0^\infty J_o(mr)J_1(ma)\left[A^i(m)m^2e^{mz}+B^i(m)m^2e^{-mz}\right. \\
&\quad \left.-C^i(m)m(1-2\nu_i-mz)e^{mz}+D^i(m)m(1-2\nu_i+mz)e^{-mz}\right]dm \\
\sigma_{rr}^i &= -qa\int_0^\infty J_o(mr)J_1(ma)\left[A^i(m)m^2e^{mz}\right. \\
&\quad +B^i(m)m^2e^{-mz}+C^i(m)m(1+2\nu_i+mz)e^{mz} \\
&\quad \left.-D^i(m)m(1+2\nu_i-mz)e^{-mz}\right]dm+qa\int_0^\infty \frac{J_1(mr)}{mr}J_1(ma)\left[A^i(m)m^2e^{mz}\right. \\
&\quad \left.+B^i(m)m^2e^{-mz}+C^i(m)m(1+mz)e^{mz}-D^i(m)m(1-mz)e^{-mz}\right]dm \\
\sigma_{\theta\theta}^i &= -qa\int_0^\infty \frac{J_1(mr)}{mr}J_1(ma)\left[A^i(m)m^2e^{mz}+B^i(m)m^2e^{-mz}\right. \\
&\quad \left.+C^i(m)m(1+mz)e^{mz}-D^i(m)m(1-mz)e^{-mz}\right]dm \\
&\quad -2\nu_i qa\int_0^\infty J_o(mr)J_1(ma)\left[C^i(m)me^{mz}-D^i(m)me^{-mz}\right]dm \\
\tau_{rz}^i &= -qa\int_0^\infty J_1(mr)J_1(ma)\left[A^i(m)m^2e^{mz}-B^i(m)m^2e^{-mz}\right. \\
&\quad \left.+C^i(m)m(2\nu_i+mz)e^{mz}+D^i(m)m(2\nu_i-mz)e^{-mz}\right]dm \\
w^i &= -\frac{1+\nu_i}{E_i}qa\int_0^\infty J_o(mr)J_1(ma)\left[A^i(m)me^{mz}-B^i(m)me^{-mz}\right. \\
&\quad \left.-C^i(m)(2-4\nu_i-mz)e^{mz}+D^i(m)(2-4\nu_i+mz)e^{-mz}\right]dm \\
u^i &= \frac{1+\nu_i}{E_i}qa\int_0^\infty J_1(mr)J_1(ma)\left[A^i(m)me^{mz}+B^i(m)me^{-mz}\right. \\
&\quad \left.+C^i(m)(1+mz)e^{mz}-D^i(m)(1-mz)e^{-mz}\right]dm
\end{aligned} \tag{5.46}$$

However, one needs to know the values of the coefficients $A^i(m)i$, $B^i(m)$, $C^i(m)$, and $D^i(m)$. These are obtained from the boundary conditions. The boundary conditions for the present problem (at the surface, at interface and at infinite depth) are discussed in the following.

5.4.2 BOUNDARY CONDITIONS

At the surface, as discussed earlier, the vertical stress at the circular region will be equal to pressure applied (q), and shear stress will be equal to zero. This is already represented in the form of Equation 5.40.

For perfectly bonded interface (also called as rough interface, when the layers cannot slide over the other) between ith and $(i+1)$th layer, the σ_{zz} at the bottom of the ith layer should be equal to that of at the top of the $(i+1)$th layer. So also will be w, u and τ_{rz}. Thus, the boundary conditions will be,

$$\begin{aligned} \sigma_{zz}^{i,b} &= \sigma_{zz}^{i+1,t} \\ w^{i,b} &= w^{i+1,t} \\ \tau_{rz}^{i,b} &= \tau_{rz}^{i+1,t} \\ u^{i,b} &= u^{i+1,t} \end{aligned} \tag{5.47}$$

For perfectly smooth interface (that is, when the layers can slide over each other), the σ_{zz} and w of at the bottom of the ith layer should be equal to that at the top of the $(i+1)$th layer. However, the shear stresses for both the ith and $i+1$th should be equal to zero since sliding is allowed. Hence, the boundary condition becomes

$$\begin{aligned} \sigma_{z}^{i,b} &= \sigma_{z}^{i+1,t} \\ w^{i,b} &= w^{i+1,t} \\ \tau_{rz}^{i,b} &= 0 \\ \tau_{rz}^{i+1,t} &= 0 \end{aligned} \tag{5.48}$$

It may be mentioned that perfect smooth or perfect bonded interface condition is a theoretical idealization. In reality, a situation somewhere in between may be realized. Bond strength between to adjacent layers may be improved by texturing or by applying adhesive agent at the interface; bond strength could be low if a layer rests on over a smooth surface or if smooth separation layer is provided at the interface.

At infinite depth, one can assume that the stresses and displacements are all zero, that is,

$$\begin{aligned} \sigma_{zz} = \sigma_{rr} = \sigma_{\theta\theta} = \tau_{rz} &= 0 \\ u = w &= 0 \end{aligned} \tag{5.49}$$

This condition leads to, $A^n(m) = 0$ and $B^n(m) = 0$. If, however, there is a rigid base at the bottom of nth layer, for rough interface, one can write [133]

$$\begin{aligned} w^{n,b} &= 0 \\ u^{n,b} &= 0 \end{aligned} \tag{5.50}$$

For smooth interface between the nth layer and rigid base, one can write [133]

$$\begin{aligned} w^{n,b} &= 0 \\ \tau_{rz}^{n,b} &= 0 \end{aligned} \tag{5.51}$$

Since this is a n-layered structure, there will be $4 \times n$ number of unknowns for each value of m. An n-layered pavement has $(n-1)$ interfaces. Thus, from Equation 5.47 or 5.48 (as the case may be), one obtains $4 \times (n-1)$ number of equations. Equations 5.40, and 5.49 (or Equation 5.50 or Equation 5.51 as the case may be) provide additional four equations. Thus, a total of $4 \times n$ equations are obtained which can be used to solve $4 \times n$ of unknowns, for each value of m. Figure 5.8 shows a typical output from analysis of a multi-layered elastic structure. Since all the layers participate in load sharing, the stress or strain value of a specific point within the pavement structure will reduce if the elastic modulus or thickness of any layer of the pavement structure (irrespective of whether the layer is above or below the point) is increased.

5.4.3 DISCUSSIONS

The present ϕ function (refer Equation 5.38) can as well be used for solving the half-space problem (as discussed in Section 5.3), and one may refer to [45] for a worked out solution.

It may be noted that circular uniform loading is generally assumed for analysis of asphalt pavement idealized as a multi-layered structure. It also synchronizes well with the axi-symmetric cylindrical coordinate system assumed for the analysis. However, one can solve the problem in Cartesian coordinate system as well, and therefore, axi-symmetry condition is no longer required. Interested reader can refer to, for example, [191] for an alternative approach.

Further, a typical tyre imprint neither may look circular nor the loading be uniform (refer Figure 5.9). If the weight of the wheel is Q and the tyre contact pressure is q, then the equivalent radius (a) of the tyre imprint can be calculated as

$$a = \sqrt{\frac{Q}{q\pi}} \tag{5.52}$$

It may be recalled that the half-space formulation as well as multi-layered formulation is based on the assumption that the structure is infinity extended (along the X and Y direction, in Cartesian coordinate system). In engineering sense, this assumption may be reasonable given that the tyre imprint radius (a) is much smaller than the width of the pavement.

Instead of vertical pressure, one can also assume that a horizontal shear stress stress (q_s) is applied by the wheels (refer Figure 5.10). In that case, the surface

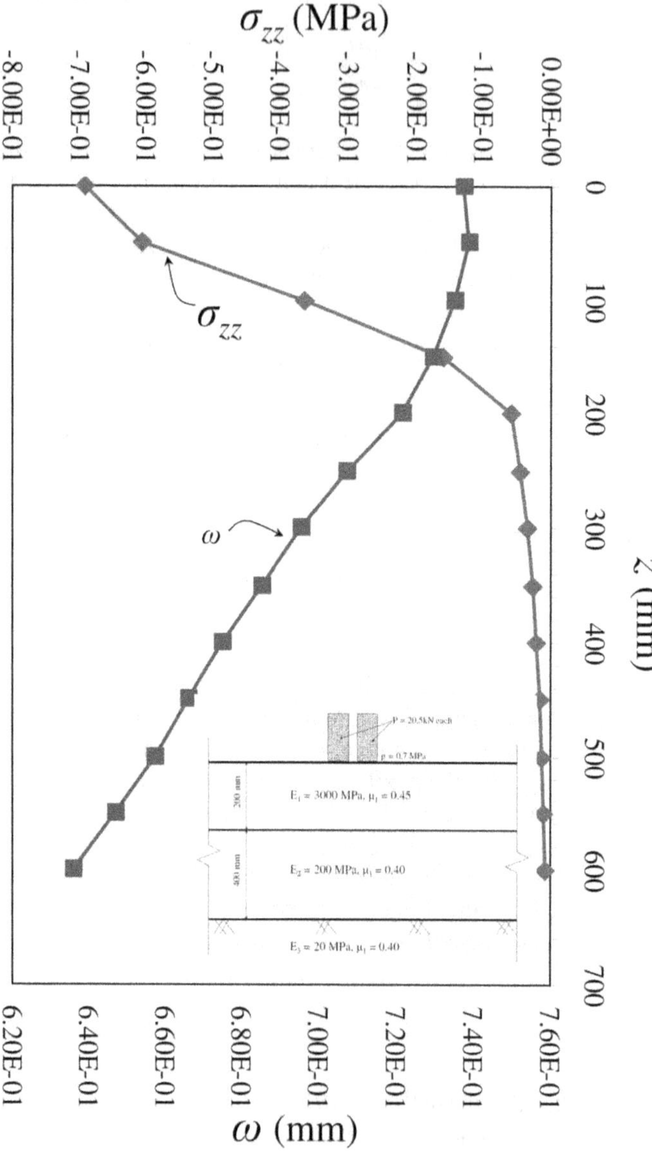

Figure 5.8 Typical output from an analysis of a multi-layered elastic structure

boundary condition becomes the following:

$$\begin{aligned} \tau_{rz}^{1,t} &= q_s \quad \text{for } 0 \le r \le a \\ &= 0 \quad \text{otherwise} \\ \sigma_{zz}^{1,t} &= 0 \end{aligned} \tag{5.53}$$

Figure 5.9 A typical tyre imprint

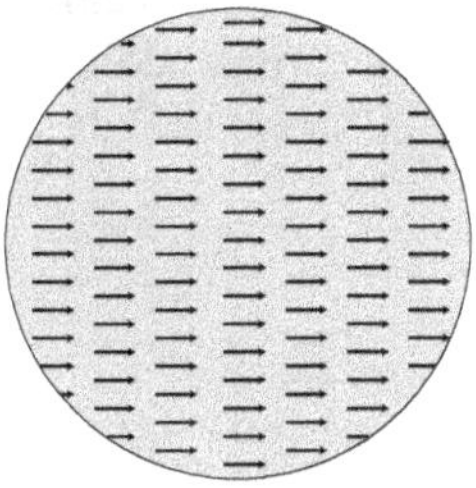

Figure 5.10 Shear stress acting within a circular area of radius a

In fact a wheel applies both, vertical pressure (q) and horizontal shear stress (q_s), and one can solve the multi-layer problem separately using Equation 5.40 and Equation 5.53, and superimpose to predict the overall effect [190].

Formulations have been suggested, for analysing the response of a multi-layered structure, when (instead of a known load) a known deformation is applied to the structure [172], when a layer has numerous cracks across the depth [173], when any of the layers is anisotropic (transversely isotropic) [43, 271], loading is non-uniform [56], or non-circular [191] and so on.

Further, it is possible to extend the n layer analysis further for a situation when some of the layers are assumed to show visco-elastic behaviour. Use of elastic-viscoelastic correspondence principle [88] (refer Equations 2.43 and 2.44) can be one of the ways of handling such problem (refer to discussions in Section 2.3.1.4). In this approach, (i) the expressions of stress-strain-displacements are taken to Lapacian domain (s), (ii) the expression of E_i can be replaced by $sE_i(s)$ (so also for ν_i, if it is assumed to be a function of time) and (iii) inversion can be performed to bring it back to time domain. A number of researchers have worked on this problem [222].

Figure 5.11 Superposition may be assumed for computing response due to multi-axle loading

Interested readers can refer to, for example [54, 152, 222], as some of the relatively recent works.

As apparent from the above discussions, a closed-form analytical solution of a n-layered elastic structure may be difficult. Thus, the results are generally obtained using numerical schemes. Further, some alternative approximate approaches have been suggested where the multi-layered problem is solved using a single layer formulation (that is, half-space), with certain additional simplifications. One can refer to [71, 268, 300] for discussions on such methods.

5.5 CLOSURE

A number of softwares/algorithms are available which performs such analyses of multi-layered structure, for example, [69, 73, 113, 128, 147, 160, 214, 225, 294, 341] and one can refer to [54, 128, 135, 290] for more discussions on such available tools. Some of the softwares/algorithms have provisions for incorporating advanced material models.

Since it is possible to analyse multi-layered structure with various material behvaiour and geometry [172] using efficient computational system in the present days, there may not be any need to estimate an equivalency factor between two layers made up of different materials, or the equivalent single layer strength (elastic modulus) of two or multiple layers, etc. Such computational power also enables one to analyse a complex arrangement of axle and wheel configuration (refer Figure 5.11). With this, assumption of load dispersion angle or equivalent single wheel load (ESWL), etc., may no longer be required.

6 Temperature Stress in Asphalt Pavement

6.1 INTRODUCTION

Change in temperature other than the temperature profile at which the pavement is stress-free (refer to the discussions in Section 4.3 in the beginning) is expected to cause thermal stress to a pavement structure. However, the asphalt mix being a rheologic material also dissipates the stress, thus developed. That is, continuous variation of thermal profile induces thermal stress to the asphalt pavement which also continuously gets dissipated. Thus, thermal stress in asphalt pavement may be negligible in moderately cold to warm places. However, in the regions with extreme cold climate, there can be significant accumulation of thermal stress within a short time-period (shorted than the time needed for its dissipation). This chapter presents a formulation to estimate the thermal stress in asphalt layer (for known rheological properties) due to any given variation of temperature.

6.2 THERMAL PROFILE

Simple formulation for estimation of thermal profile across the pavement layers has already been presented in Section 4.2. Various researchers have studied the thermal profile of asphalt pavement theoretically and experimentally [118, 315, 326]. Past studies suggest that thermal profile ($T(z)$) in the asphalt layer of an asphalt pavement is generally nonlinear [76].

6.3 THERMAL STRESS IN ASPHALT PAVEMENT

For viscoelastic material, the thermal stress dissipates over time. If temperature T would not have varied with time (t), the thermal stress (for fully restrained condition) in the asphalt layer can be calculated (refer Equation 4.11) as follows:

$$\sigma_{yy}^{T}(t) = \sigma^{T}(t) = -E_{rel}^{T}(t)\alpha(T - T_o) \tag{6.1}$$

where $E_{rel}(t)$ is the relaxation modulus of the asphlatic material. The $\sigma^T(t)$ will gradually decrease even if the temperature remains constant over time. Further, $\sigma^T(t)$, will be higher if the temperature is low. This has been schematically presented in Figure 6.1.

However, the temperature, in a pavement structure, does not remain constant, the temperature ($T(z,t)$) varies both with depth (z) and time (t) (refer Section 4.2). Such a variation will affect the restrained strain (refer Equation4.10), which in turn will

DOI: 10.1201/9781003190769-6

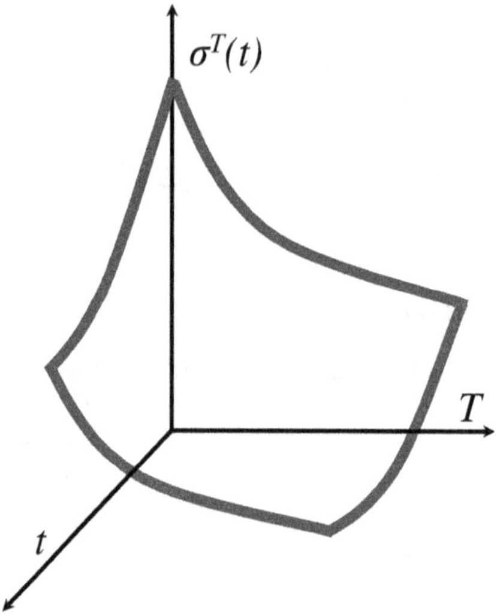

Figure 6.1 A conceptual diagram of variation of thermal stress ($\sigma^T(t)$) with temperature (T) and time (t)

influence $\sigma^T(z,t)$. Asphalt mix being rhelogical material, $\sigma^T(z,t)$, will be affected by the history of the temperature variation as well. Thus, the considerations involved in deriving an expression for $\sigma^T(z,t)$ is presented in the following:

- Assuming asphalt mix is a linear viscoelastic material, the thermal stress, can be determined by using Boltzmann superposition principle (refer Equation 2.41), as follows [51, 121, 197, 198, 222, 235, 265, 265]:

$$\sigma^T(t) = \int_0^t E_{rel}^{T(t)}(t-\zeta)\frac{\partial \varepsilon(\zeta)}{\partial \zeta}d\zeta \tag{6.2}$$

  where ζ = dummy variable for time.
- Using Equation 4.10, one can write [197, 235]

$$\frac{\partial \varepsilon(\zeta)}{\partial \zeta} = -\alpha\frac{\partial T}{\partial \zeta} \tag{6.3}$$

  where α = coefficient of thermal expansion (of asphalt in this case). It is assumed that α does not vary with temperature.
- It may be noted that since temperature is constantly changing, the $E_{rel}^{T(t)}$ needs to be converted to a equivalent value $E_{rel}^{T_r}$ at some standard temperature [198, 222, 235, 265], where T_r is a reference temperature. This can be done by invoking the time-temperature superposition principle, and assuming asphalt is a thermorheologically simple material [265].

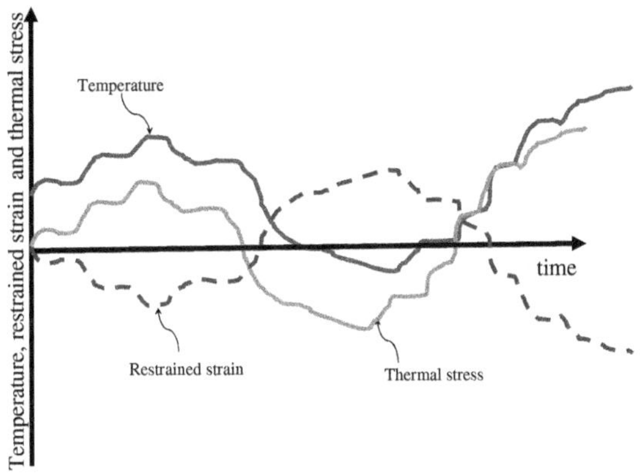

Figure 6.2 Schematic representation of variation of temperature, restrained strain and viscoelastic stress in asphalt layer

Using Equation 2.53 and considering that the reduced time at a given time $(t-\zeta)$ is sum of all the reduced times up to this time [88,198,222,265], one can derive the function for calculation of reduced time as follows [222,235]:

$$f(t-\zeta) = \int_0^{(t-\zeta)} \frac{dt''}{\alpha_T} \tag{6.4}$$

where t'' = dummy variable representing time. Suitable expression for α_T can be chosen in the form of Equation 2.55 or Equation 2.56. It is assumed that the reduced time is calculated with reference to some standard temperature, T_r. It may be noted that α_T is a function of T_r and $T(t'')$, where $T(t'')$ represents the temperature of the asphalt layer (at a given z) which is constantly varying with time, t''.

- The above considerations are assumed hold independently for each value of z.

Considering the above, the thermal stress in asphalt layer at any t and z can be expressed as [198,214,222,235,265]

$$\sigma^T(z,t) = -\alpha \int_0^t E_{rel}^{T_r}(z, f(t-\zeta)) \frac{\partial T(z,\zeta)}{\partial \zeta} d\zeta \tag{6.5}$$

Equation 6.5 can be used for calculation of thermal stress within the asphalt layer at any time and depth. One can choose a suitable rheologic model for asphalt material (refer to the discussions in Section 2.3.1) and numerically obtain the estimated value of $\sigma^T(z,t)$. One can refer to, for example, [235,265] for such a study on single layer, [222] for a study on multiple layers. A typical expected variation of temperature, restrained strain and viscoelastic stress is schematically shown in Figure 6.2.

6.4 CLOSURE

$\sigma^T(z,t)$ is a function of both the temperature as well as rate of drop of temperature (refer Equation 6.5). For low temperature and high rate of cooling, the stress developed will be high. If the drop of temperature is slow, asphalt layer may get time to dissipate the thermal stress. If the thermal stress developed is more than the tensile strength of the material (at that z and t), thermal crack may originate [83, 122]. The issues related to estimation of thermal crack spacing have been taken up in Chapter 8.

7 Finite Element Analysis of Pavement Structures

7.1 INTRODUCTION

This chapter discusses the applications of finite element method (FEM) for analysis of pavement structures. FEM is used for computationally solving various boundary value problems (for example, problems related to structural stress analysis, heat and mass transfer, fluid flow, electric and magnetic field and so on) which can typically be expressed in the form of differential equations. For structural analysis (for example, a pavement structure), application of FEM becomes useful when the there are complexities associated with the geometry, loading configuration or material properties (say, heterogeneity, non-linearity, anisotropy, etc.), where it may be difficult to reach closed-form solutions. The FEM converts the boundary value problem into a system of linear equations as its numerical equivalent.

7.2 BASIC PRINCIPLES

For structural analysis problem, the objective is to find out the displacement field, in terms of translation or rotation along the three axes (depending on the degrees of freedom chosen) within the structure.

The FEM approach for the solution of a typical structural problem (static and linear) involves the following steps:

- The structure is discretization (also called as mesh generation) into a number of elements of finite sizes. Depending on the nature of problem, the elements could be one (line elements), two (say, triangular or quadrilateral elements) or three dimensional (say, brick, tetrahedron elements), with predefined degrees of freedom. The elements are connected with other elements at the nodes.
- For any given element, one can write

$$\{q\} = [k]\{u'\} \tag{7.1}$$

  where $\{q\}$ is the nodal force matrix of the element, $[k]$ is the element stiffness matrix and $\{u'\}$ is the nodal displacement matrix of an element. $[k]$ is a symmetric (for linear elastic materials) square matrix, and the number of rows (or columns) are equal to the degrees of freedom that the element is assumed to have.

  There are various methods to calculate the stiffness matrix of an element. These can be based on direct method (also known as the strong form), or, variational method or weighted residual method (for example, Galerkin's

DOI: 10.1201/9781003190769-7

method, collocation method, least square method, etc.) (also known as weak form). These are briefly discussed in the following.

- Direct method: For simpler structural configuration, the stiffness matrix can be obtained directly form the equilibrium condition. For example,
  - For a linear spring (with spring constant k_s) as a line element, the stiffness matrix will be $[k] = \begin{bmatrix} k_s & -k_s \\ -k_s & k_s \end{bmatrix}$

  In case the governing equation is expressed in differential equation, and closed-form analytical solution is available, then the displacement field can be obtained directly.
- Variational method: It is assumed that the minimization of the total potential energy of the system provides the possible deflected shape (due to kinematicaly admissible displacement field) of the structure. Variational calculus is utilized in this approach.

  The total strain energy in a structural system (over the volume Ω) is given as

$$U_s = \frac{1}{2}\int_{\Omega} \{\varepsilon\}^T \{\sigma\} d\Omega \tag{7.2}$$

Using Equations 1.19 and 1.23, one can write

$$U_s = \frac{1}{2}\int_{\Omega} [[B]\{u\}]^T [C][B]\{u\} d\Omega \tag{7.3}$$

The potential energy due to body force (F_Ω) and surface traction (F_Γ) is

$$U_p = -\int_{\Omega} \{u\}^T F_\Omega d\Omega - \int_{\Gamma} \{u\}^T F_\Gamma d\Gamma \tag{7.4}$$

The total potential energy of the deformed body is therefore,

$$U = U_s + U_p \tag{7.5}$$

For minimization of total potential energy,

$$\frac{\partial U}{\partial \{u\}} = 0 \tag{7.6}$$

This finally gives

$$[k] = \int_{\Omega} [B]^T [C][B] d\Omega \tag{7.7}$$

- Weighted residual methods: Galarkin's method is one of the weighted residual methods. In this method, an approximate solution is assumed to the governing differential equation formulated. Thus, for 1-D case,

$$u(x) = \sum_{i=1}^{n} c_i F_i(x) \tag{7.8}$$

where c_i are the coefficients and $F_i(x)$ are the proposed trial functions. Since this is an approximate solution, by putting this solution in the differential equation will give rise to the residual, $R(x)$. The unknown coefficients are obtained by minimizing the residuals as follows:

$$\int R(x)F_i(x)dx = 0 \text{ for } \forall i \tag{7.9}$$

- Global stiffness matrix $[K]$ is formed assembling the local (elemental) stiffness matrices, keeping in view the nodes that the adjacent elements share, their compatibility and the coordinate system. For example,
  - If two springs with spring constants k'_s and k''_s are connected in series, then assuming each spring as line element with single degree of freedom per node, $[K]$ can be written as

$$[K] = \begin{bmatrix} k'_s & -k'_s & 0 \\ -k'_s & k'_s + k''_s & -k''_s \\ 0 & -k''_s & k''_s \end{bmatrix}$$

  since there are three nodes in the system, and each node having one degree of freedom, the total degree of freedom is three.
  - Similarly, if three springs with spring constants k'_s, k''_s and k'''_s are connected in series, and assuming each spring as line element with single degree of freedom per node, $[K]$ can be written as

$$[K] = \begin{bmatrix} k'_s & -k'_s & 0 & 0 \\ -k'_s & k'_s + k''_s & -k''_s & 0 \\ 0 & -k''_s & k''_s + k'''_s & -k'''_s \\ 0 & 0' & -k'''_s & k'''_s \end{bmatrix}$$

In case the alignment of the element coordination system is different than the coordinate system assumed for the system, then the elemental stiffness matrices need to be suitably multiplied with the rotation matrix $[R]$ before assembling to form the global stiffness matrix.
Global stiffness matrix is a square matrix with number of rows (or number of columns) equal to the total number of degrees of freedom of the system. Thus, the equation of the system can be written as

$$\{Q\} = [K]\{U'\} \tag{7.10}$$

where $\{Q\}$ is the nodal force matrix of an system and $\{U'\}$ is the nodal displacement matrix of the system.

- After incorporating the boundary conditions (the known displacements at the specific nodes, including restraint in displacement), and the known forces, the system of equations is solved so as to find the displacements at different nodes[1].
- Once the displacements at the nodes are known $\{u'\}$, the displacements field at any other location within the element $\{u\}$ can be obtained through piece-wise interpolation. It may be written as

$$\{u\} = [N]\{u'\} \tag{7.11}$$

where $[N]$ contains the shape function terms. For simplicity, polynomial functions are generally chosen. For a given element type (say, triangular element or brick element), the specific shape function coefficients are derived.
- From the known displacement field, the strains can be estimated (refer to Equation 1.19), and therefrom the stresses can be estimated (refer to Equation 1.23) at any points of interest.

The basic principles for FEM have been presented above in brief. Although the above discussions are centred around structural analysis, where the deflection field is the unknown, similar concept can be utilized for FEM solution for estimation of heat or fluid flow, electric or magnetic potential. One can refer to text books (for example, [25, 89, 241]) for detailed treatise on FEM.

7.3 FEM ANALYSIS FOR PAVEMENT STRUCTURES

As discussed through the previous chapters, closed form solutions towards analyses of asphalt and concrete pavements can only be developed for limited cases. The FEM approach becomes very useful to take care of various forms of complexities (non-linearity, non-uniformity, asymmetry, discontinuity and so on) in geometry, material property and loading, once the governing differential equation is framed. Accordingly the researchers have considered 2-D or 3-D models with different element types, pavement boundary conditions, interface contact types and materials with various constitutive relationships.

Validation of FEM results is an important aspect. Analytical solution is available only for limited and idealized scenarios. Laboratory or field validation involves careful consideration so as to re-create the geometry, loading and other physical conditions assumed in the FEM model.

Figure 7.1 shows typical boundary conditions of pavement structure in axi-symmetric configuration for FEM implementation, and Figure 7.2 shows typical mesh generation for pavement structure for FEM implementation

A large pool of research studies is available dealing with FEM application to solve stress–strain due to static and dynamic loading, thermal stress and heat

[1]The numerical solution requires special care given that $[K]$ generally tends to be a sparse and banded matrix

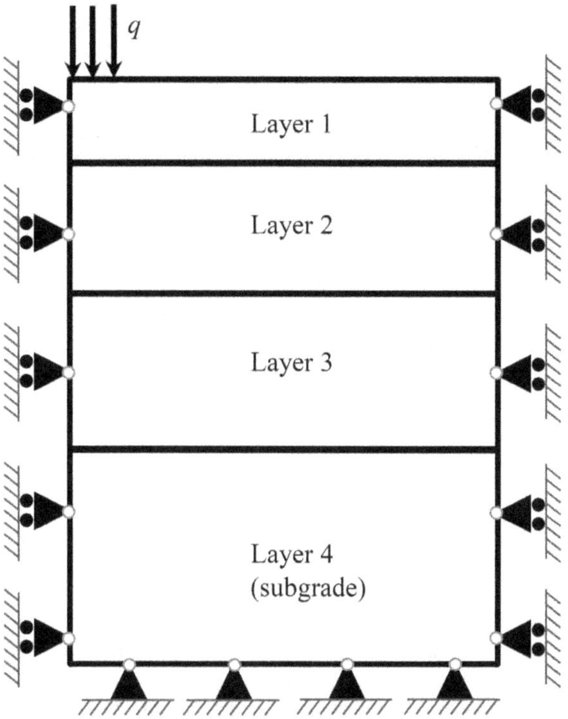

Figure 7.1 Typical boundary conditions of pavement structure in axi-symmetric configuration for FEM implementation

flow analysis, pavement-tyre interaction, flow of water through porous pavements, stresses due to moisture gradient etc. Studies also have been done to predict the responses of individual materials.

With the increased computational power and availability of general purpose FEM softwares and graphical user interface the complexities associated with mesh generation, assembling stiffness matrix, solving large-sized sparse and banded matrix, pre and post processing can be handled with greater efficiency and accuracy. A number of pavement specific analysis softwares are also available - for example, ILLI-SLAB, ILLI-PAVE (2D), MICH-PAVE (2D), KENSLAB, EverFE and so on [72, 113, 128, 294]. A few works related to FEM implementation on pavement engineering problems are summarized in the following.

- FEM implementation has been done to computationally obtain results for the models developed for pavement materials, for example, micromechanical model of asphalt mix [14, 22, 155, 250], damage of asphalt mix-related loading and moisture [63, 263].
- Related to concrete pavements, studies have been done to obtain response as beam resting on elastic foundation [218], slab with dynamic loading [255], distribution of thermal stress in doweled concrete pavement [185], etc.

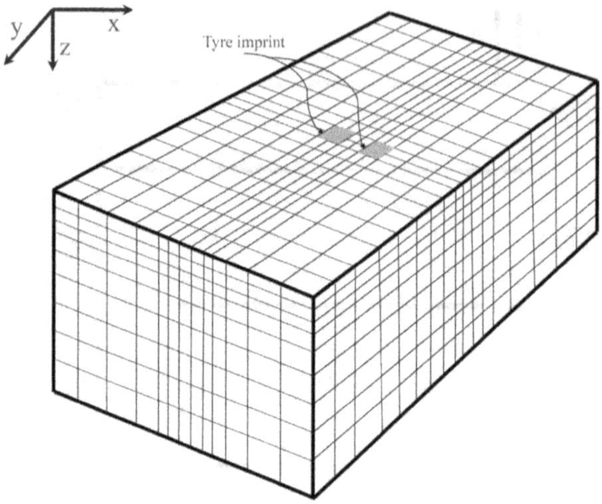

Figure 7.2 Typical mesh generation for pavement structure for FEM implementation

- Related to asphalt pavements, studies have been done to obtain response due to non-uniform tyre contact pressure, different shapes of tyre imprint and tyre geometry [140,262]. Various studies include prediction of response due to various loading amplitude, speed, axle configuration and interface conditions [254, 329]. Researchers have studied dynamic response of asphalt pavement modelled as multi-layered media [31, 80, 254, 324, 324], including response of pavement due to movement of traffic speed deflectometer [196],
  FEM implementation also has been done to predict long-term response of asphalt pavement considering the tyre-pavemnt interaction [29], prediction of various distresses (for example, rutting [9], crack initiation and propagation [82]) and so on. FEM models have been developed to study the interaction between aggregates in pavement surface, and the tyre [12, 103].

7.4 CLOSURE

With the rapid progress of computational power and development of advanced modelling, researchers are able to simulate results with improved accuracy and closer to the realistic scenario. This will eventually help to augment the pavement design process (principles discussed in Chapter 8) with the precise incorporation of incremental damage and their locations [100,262] due to repetitive loading and variations of environmental conditions.

8 Pavement Design

8.1 INTRODUCTION

A pavement design may involve structural, functional and drainage design. This chapter deals with the basic principles of structural design of a pavement. The goal of this chapter is to establish a link between the analysis schemes developed in the previous chapters with the current approaches of pavement design. A large pool documents are available in design of pavement structure, in the form of text books [101, 128, 195, 221, 328], codes/guidelines [1, 91, 135, 136, 207, 214, 220, 224, 225, 273, 292, 295] and various background papers [62, 268, 269, 291], and interested readers can refer to these for further study. The discussions in this chapter, therefore, have been kept brief.

Structural design of pavement primarily involves estimation of thickness. The thicknesses are so provided so that the pavement is able to survive against the structural distresses, caused due to traffic loading and effect of environment. It may be reminded that there can be various other distresses which are non-structural in nature. Some of the primary modes of structural distresses are load fatigue, thermal fatigue, rutting, low temperature shrinkage cracking, top-down cracking, punchouts (generally relevant for continuously reinforced cement concrete pavements), crushing, etc. One can refer to, for example [208, 273], for an overview of various types of distresses that may occur to a pavement structure.

For asphalt pavement, the structural design generally involves estimation of the thicknesses of base, sub-base and surfacing. For concrete pavement, structural design generally involves estimation of thicknesses of base and the concrete slab; for concrete pavements, it also includes estimation of joint spacing and detailing (spacing, diameter and length) of dowel and tie bars.

In principle, there is not much of a difference between the pavement design approaches for a highway pavement and a runway/taxiway pavement [87, 126, 220] or a dockyard pavement – the input parameters may be different (for example, load application time in dockyard [272] may be longer than taxiway, the lateral wander of wheels in runway may be distributed over a larger length than in highways and so on), but the design philosophy (in terms of mechanistic-empirical design approach), in general, remain the same.

8.2 DESIGN PHILOSOPHY

Historically, a number of approaches have been suggested for designing the pavement structure, ranging from empirical, semi-empirical to mechanistic-empirical method. Some of the earlier approaches were based on (i) experience, (ii) bearing capacity, (iii) shear strength, (iv) deflection and so on. In some of these methods, the layer thickness values are so designed that the maximum value (of bearing stress or

DOI: 10.1201/9781003190769-8

shear stress or deflection) does not exceed the strength (say, bearing strength or shear strength of the material) or the limiting criterion (say, deflection). Certain initial design approaches did not originally include traffic repetitions as a parameter [328]. However, it was realized that a pavement structure does not necessarily fail due to single application of load due to ultimate load bearing conditions (although there are some exceptions, as discussed towards the end of Section 8.4.1); rather, it is the repetitions of load (or environmental variations) that cause failure. Repetitions gradually emerged as one of the important considerations in the pavement design process.

One can refer to various other books/papers, for example, [128, 211, 300, 325], etc., for a brief review of the historical perspective of pavement design[1].

In the mechanistic-empirical pavement design approach, mechanistically estimated initial stress–strain values at critical locations are empirically related to the cumulative repetitions for individual distresses. Such relations are also known as transfer functions. For example, horizontal tensile stress/strain at the bottom of any bound layer can be related to fatigue life [224, 225, 273, 295], vertical compressive strain(s) on individual layers [224, 225, 273, 295] or shear stress or principal stress can be related to permanent deformation (rutting) [273], vertical compressive stress can be related to crushing of cemented layer [273] and so on. Some of the initial works which formed the basis of mechanistic-empirical method are due to [78, 212, 247]. Now, this approach is quite widely being used in various countries and a number of guidelines/codes follow mechanistic-empirical method of pavement design [1, 68, 91, 207, 214, 224, 225, 273, 292, 295]. Some of the structural distresses considered in mechanistic-empirical pavement design are discussed further in the following:

8.2.1 LOAD FATIGUE

Repetitive application of load causes load fatigue failure to the bound layer of the pavement. In the laboratory this is simulated by applying repetitive flexural loading on beams of various geometry (refer to Section 2.3.3). Since the maximum tensile stress or strain occurs at the bottom of bound layer, crack initiates there and propagates upwards as repetitions progresses, hence, this is known as bottom up cracking. The laboratory conditions of fatigue testing being different than field (for example, differences in loading pattern, rest period, temperature, boundary conditions of the sample etc.), calibration/adjustment is needed on the laboratory equation to make it usable as a design equation [11, 297]. Further, the definitions of fatigue failure in laboratory (refer to the discussions in Section 2.3.3) may be different than that of in the field (which may be based on percentage appearance of characteristic surface crack). One can refer to, for example, [19, 270, 276] for a brief overview of the various load fatigue transfer functions used for pavement design purpose.

[1] It must, however, be pointed out that some of the earlier design approaches may still be relevant, for example, deflection is used as a criteria for overlay design [17, 137, 138, 225], bearing capacity can be used to evaluate stress and bearing capacity of individual layers [291], shear strength can be used as a criteria in integrated approach for mix- design-pavement-design [98] and so on.

8.2.2 RUTTING

Permanent deformation in pavement along the most traversed wheel path is called as rutting; rutting is measured at the surface of the pavement. Rutting can happen due to (i) compaction (reduction) of air-voids and/or (ii) shear flow of the material. Since rutting is measured at surface, the contribution may come from one or more layer(s) due to either of these mechanisms. Figure 8.1 schematically shows the rutting contributed by the *i*th layer due to compaction (Figure 8.1(a)) and shear flow (Figure 8.1(b)).

Various empirical, semi-empirical or mechanistic (viscoelastic/viscoplastic) models have been suggested where elastic strain(s) of pavement layer(s), pavement thickness(es), temperature, asphalt content, air-voids, dynamic modulus, resilient modulus, aggregate gradation, traffic repetitions, moisture content, state of stress, rheological parameters, etc., have been related (mechanistically or empirically) to predict the rut depth of asphalt layer or granular layer or overall rutting [9, 55, 129,

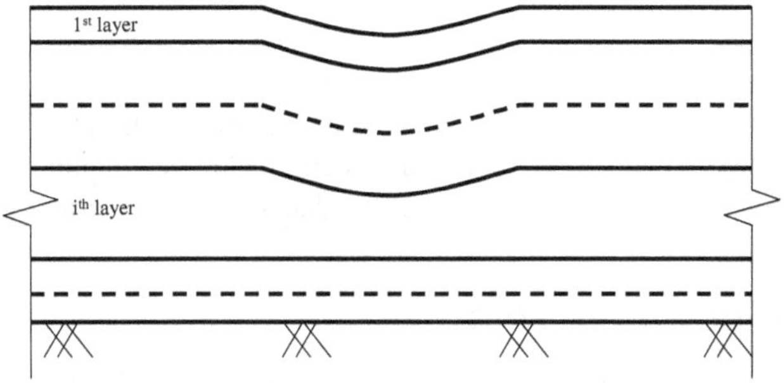

(a) Rutting of *i*th layer due to compaction

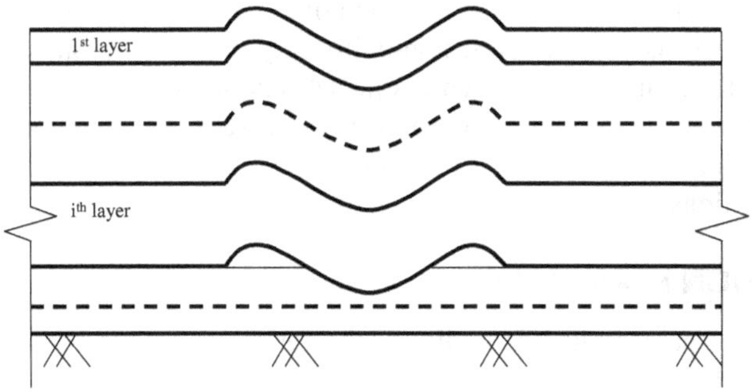

(b) Rutting of *i*th layer due to shear movement

Figure 8.1 Schematic diagram representing two possible mechanisms of rutting

182, 214, 274, 342]. One can, for example, refer to [233] for a review on rutting in asphalt layer, and [171, 299] for a review on rutting in granular layer, and [321] for a study on the relative contribution of different pavement layers towards rutting.

8.2.3 LOW-TEMPERATURE SHRINKAGE CRACKING

Low-temperature shrinkage cracks along the transverse direction of the road. Such cracks are prevalent in the roads of the colder region [197]. It originates when the tensile stress generated within the bound layer exceeds the tensile strength of the material [83, 122]. Basic formulations for estimation of thermal stress have been developed in Sections 4.3.2.1 and 6.3 for cement concrete and asphalt pavement respectively. For similar environmental conditions and mix composition, the transverse shrinkage cracks are expected to be spaced equally [235, 264, 265]. The design considerations for estimating the crack spacing have been presented at a later section in this chapter.

8.2.4 THERMAL FATIGUE

Temperature variations cause alternative expansion and contraction to the pavement materials. This causes damage (to the bound layers) due to thermal fatigue which keeps on accumulating due to repetitions of the thermal cycles [10, 28, 83, 312]. Past researches suggest that variation of temperature, thickness of bound layer, maximum thermal shrinkage stress level etc. affect the damage due to thermal fatigue [10, 28, 234].

8.2.5 TOP-DOWN CRACKING

Unlike load fatigue cracking, these cracks are seen to start from top and propagate downward, for asphalt [246] or concrete [33] pavements. Conventional load fatigue theory cannot explain this since this assumes that tensile strain causes fatigue cracks to initiate, whereas the top portion of the pavement generally belongs to compression zone [48]. It is postulated that the initiation of cracks at the top may be attributed to shear stress applied by the tyres, non-uniformity in the tyre contact pressure, tensile stress at the surface generated due to wheels or axle placement, hardening of asphalt material due to aging, mix segregation, low-temperature shrinkage crack and so on [48, 246, 278]. Once initiated, it is assumed that the cracks propagate with the traffic repetitions.

8.3 DESIGN PARAMETERS

The following paragraphs briefly discuss the various parameters associated with the pavement design.

8.3.1 MATERIAL PARAMETERS

The road materials are characterized through various tests, and the material parameters are subsequently used in pavement analysis as input. Road material characterization has been covered in Chapter 2.

8.3.2 TRAFFIC PARAMETERS AND DESIGN PERIOD

Traffic parameters are used to predict the cumulative traffic count during the design period. For new road construction, traffic needs to be predicted where there exists no traffic. This is generally done considering traffic volume of existing stretches located in similar activity zone. The traffic parameters include traffic volume and its variation, axle load distribution, axle and the wheel configuration, tyre contact pressure, lateral wander of wheel, traffic growth rate, lane distribution and so on. This cumulative traffic is generally expressed in terms of million standard axle load repetitions using various empirical or theoretical equivalency factors [1, 110, 128, 214, 225, 273, 292, 295].

8.3.3 ENVIRONMENTAL PARAMETERS

Variation of moisture and temperature may affect the layer modulus values, which in turn affect the critical stress/strain values for multi-layered structure. The incremental damage for same traffic repetition may be different due the effect of environmental variations. Advanced environmental models simulate the cumulative damage considering the effect of variation of environmental parameters [207, 214].

8.4 DESIGN PROCESS

The basic formulation for estimation of stresses due to load has been presented in Chapters 3 and 5. The basic formulation for estimation of stresses due to temperature has been presented in Chapters 4 and 6. The principle of analysis of pavement using finite element method has presented in Chapter 7. The codal provisions may vary from one guideline to the other on whether and how the load and thermal stresses should be combined and considered in the design process.

The thermal stress in asphalt pavement at higher temperature is generally neglected because of its quick dissipation. The thermal stress at low temperature in asphalt pavement is used for the prediction of thermal shrinkage crack spacing (refer Section 8.4.3.1).

The thermal stress in concrete pavement is highest at the interior portion (maximum restraint) and least at the corner (least restraint). Further, the thermal stress is tensile at the bottom of the slab during day time (when, $T_t > T_b$) and is compressive at the bottom during the nighttime (when, $T_t < T_b$) (refer Section 4.3.1.6). Thus, the load stress and thermal stress may be additive or subtractive depending on the location and the thermal profile. A schematic diagram indicating relative magnitudes (not to scale) and the nature (tension/compression) of load and thermal stresses (typically bending) in concrete pavement has been presented in Figure 8.2. In the diagram edge stress due to load means the maximum stress is at the edge due to wheels positioned near the edge and so on (also refer to Section 3.4). Corner, edge and interior are generally the three critical positions of concrete pavement slab (refer Figure 8.4). The load and the thermal stress (and their overall effects) need to be analysed individually to identify the scenario that would finally govern the design.

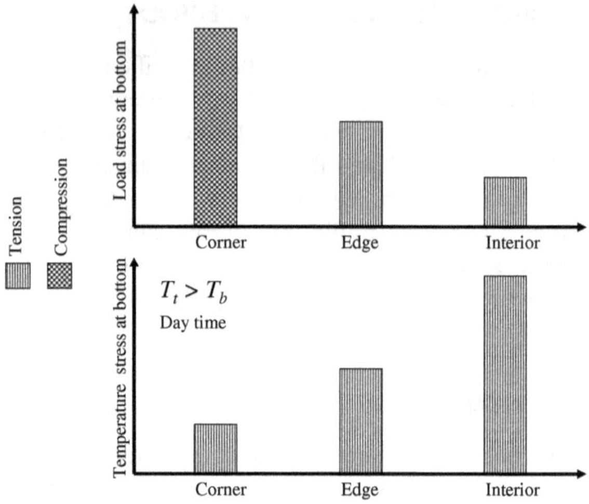

Figure 8.2 A schematic diagram indicating possible relative magnitudes (not to scale) and nature (tension/compression) of load and thermal bending stresses in concrete pavement at corner, edge and interior

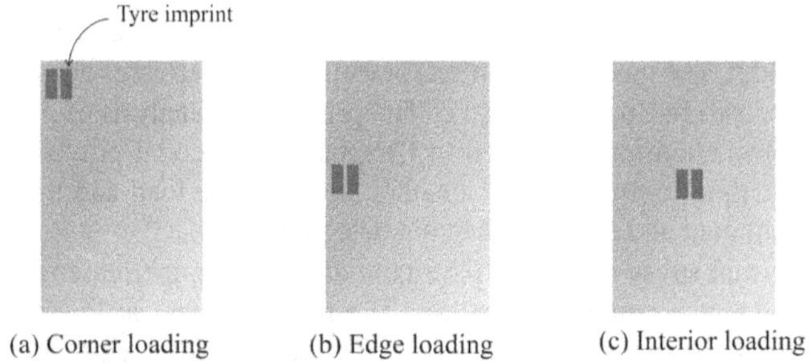

Figure 8.3 Corner, edge and interior as the critical positions of concrete pavement slab

Further, existence of moisture gradient may also affect the overall stress value. However, generally its effect on stress is opposite to that of due to temperature [16]. Hence, considering only the bending stress due to load and temperature (and ignoring the stress due to moisture gradient) may make the design conservative.

8.4.1 THICKNESS DESIGN

8.4.1.1 New Pavement

Figure 8.4 shows a generic pavement thickness design scheme. The predicted cumulative traffic is used in the transfer functions to obtain the allowable values of critical

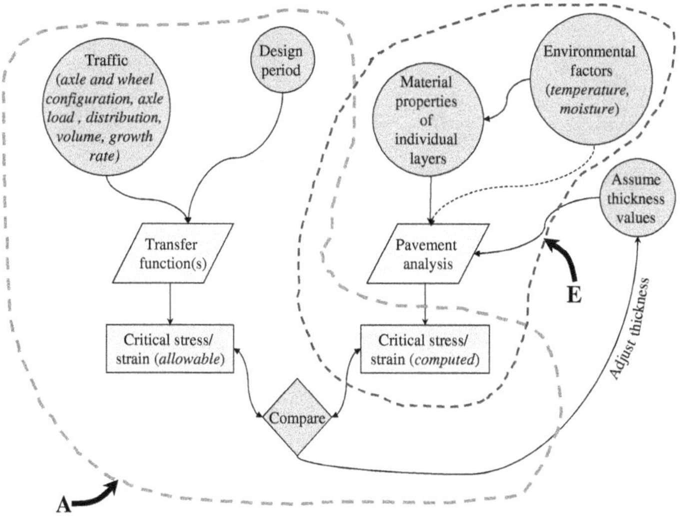

Figure 8.4 A generic pavement design scheme, where thicknesses are decided based on critical stresses/strains

stress/strain for a given type of structural distress. Further, from the assumed values of the thicknesses and the material properties, environmental parameters, and for a standard loading configuration, the stress/strain at the critical locations are computed using pavement analysis. These computed stress/strain values are compared with the corresponding allowable values. The thicknesses are adjusted iteratively until the values become approximately equal, and subsequently the design is finalized.

As per Figure 8.4, the iteration for design thicknesses is based on comparison between the allowable and computed critical stress/strain values. Alternatively, comparison can be also performed in terms of the traffic repetitions, that is, comparison between the number of expected traffic repetitions (T) and number of traffic repetitions the pavement can sustain (N). Thus, the portion marked 'A' of Figure 8.4 can be changed to develop Figure 8.5, as an alternative design scheme. The value of N for a given distress can be obtained by putting the critical stress/strain values to the corresponding transfer function.

For a deterministic design, a designer tries to adjust the thicknesses so that $N \approxeq T$. That means,

$$\frac{T}{N} \approxeq 1 \tag{8.1}$$

Given that there are seasonal variations (for example, variation of temperature affects modulus of asphalt layer, variation of moisture affects modulus of subgrade etc), variations in axle load (different axle load results in different strain levels) during the design period, one may like to divide the design period in smaller time periods and calculate the fractional damage (due to different phases of environmental

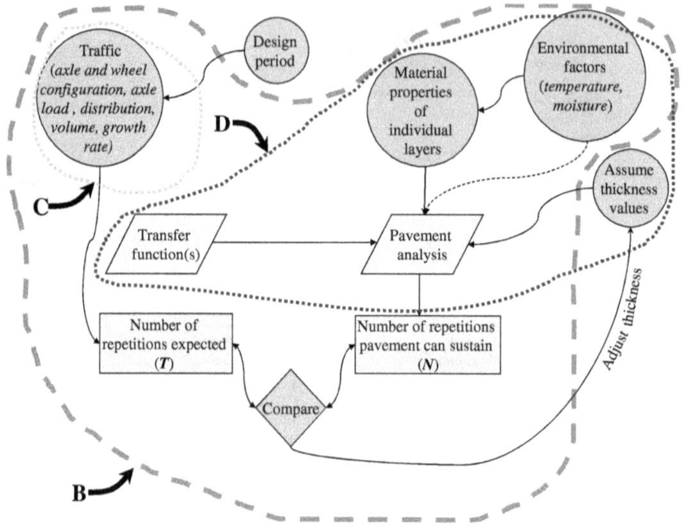

Figure 8.5 A generic pavement design scheme where thicknesses are decided based on number of repetitions

variations [295], different axle load groups [252], etc.) for each of these time period. Smaller time periods may be constituted with groups of seasons, months, days or even hours. For example, stresses may be different (say) in summer and winter months due to differences in the strength, interface and restraint conditions, further on a given month the daytime and nighttime stresses in pavement may differ, and one may like to consider these separately. Assuming fractional damages are linearly additive (refer Equation 2.64 and discussions thereof), one can write

$$\sum_{\text{EC}} \sum_{\text{AL}} \frac{T}{N} \approxeq 1 \tag{8.2}$$

where EC = different groups of environmental conditions, AL = different groups of axle loads. Thus, the portion marked as 'B' in Figure 8.5 undergoes a loop (for calculation of fractional damages for different axle load, seasons, months, days etc., and one can have as many such summations), before the design thicknesses are finalized.

In the above example, the damage (as per Equations 8.1 or 8.2) is made equal to 'one' for design purpose. That is, expected traffic repetitions (T) are made equal to the traffic repetitions the pavement can sustain (N). One may, however, decide to do a conservative design taking into account the uncertainties and inherent variabilities in various associated parameters [128, 192, 275, 276, 328]. This can be done by increasing the design traffic by a factor. Reliability analysis provides a basis of deciding this factor for a given reliability level of the pavement design. The reliability analysis for pavement design has been briefly dealt in Section 8.4.2.

8.4.1.2 Existing Pavement (Overlay)

The same principle of pavement design (discussed above) holds for design of overlay which involves provision of additional layer over the existing pavement. In this case, the pavement structure including the proposed overlay needs to be analysed to obtain the stress/strain parameters at the critical locations. For this analysis, one needs to consider the present strength (say, modulus) of the existing pavement, which might have deteriorated over the passage time (due to traffic repetitions and temperature variations). The present strength of the pavement layers at a given time during the service can be experimentally evaluated through structural evaluation at that time. Depending on the type of equipment used in the evaluation process, it may involve solution handling inverse problems (refer Section 9.4 for discussion on inverse problem). Besides the mechanistic-empirical approach for overlay design, other approaches also have been suggested. For example:

- In deflection based approach, the overlay is recommended (based on the experience gained from past studies) if the surface deflection (caused by some static or impact loading) is observed to be beyond some specified permissible limit and subsequently the overlay thickness is estimated from the observed deflection [137, 225].
- In effective thickness approach (or, equivalent thickness approach) [1, 17, 214], the thickness of the pavement is expressed in terms of an equivalent thickness considering the contributions from individual layers [1, 87]. The overlay thickness to extend the pavement life by a design traffic can be estimated as follows:

$$h_o = h'_o - kh_{ex} \tag{8.3}$$

  where h_o = overlay thickness to be provided for a given design traffic, h'_o = equivalent pavement thickness to be provided had it been designed as a new pavement, h_{ex} = equivalent thickness of the existing pavement at present, k = a factor less than one, which takes care of the fact that the existing pavement has undergone some deterioration and the equivalent thickness of the existing pavement is effectively less than what it physically is. If an asphalt overlay is to be provided on concrete pavement, or a concrete overlay is to be provided on asphalt pavement, then the Equation 8.3 can be modified as follows:

$$h_o = C\left((h'_o - k(h_{ex})^m\right) \tag{8.4}$$

  C = empirical thickness conversion factor from one type material to other (say, from asphaltic to concrete or vice versa), m = an exponent depending on the bonding between the overlay and its underlying layer.

  The formulae used for estimation of overlay thickness in different guidelines may be different (than Equations 8.3 and 8.4), but these are generally based on the above conceptualization [1, 214, 220].

It can be shown shown that the above alternative approaches of overlay design can as well be derived using the mechanistic-empirical design principle.

Some distresses may not be related to critical stress strain parameters, rather they may be related to some other parameter (for example, erosion damage has been related to energy needed to deform the slab), which in turn is dependent on traffic repetitions [214, 292]. One can still calculate cumulative damage due to traffic repetitions in this case. Therefore, the schematic diagram presented as Figure 8.5 still holds.

Some distresses, for example, crushing of materials or concrete slab punchout may be independent of the traffic repetitions. Further, one may like to check whether the sum of maximum thermal and load stress for concrete pavement should always be less than the modulus of rupture of concrete. Such a check is independent of traffic repetitions. Thermal shrinkage crack occurs at the instant when the generated tensile stress exceeds the tensile strength of the material. Thus, it also can be considered somewhat independent of traffic or thermal loading cycle. If the pavement material is a rheological material (like, asphalt mix), there will be historical effect of thermal variations causing accumulation and dissipation of stress as discussed in Chapter 6 – thus it may related to thermal cycles.

8.4.2 RELIABILITY ISSUES IN PAVEMENT DESIGN

Reliability of a pavement is the probability that it survives within the design period. In Section 8.4.1, estimation of N and T and their participation in pavement design has been explained. Deterministically, these parameters assume fixed values. However, variabilities in the parameters marked in the portion 'C' in Figure 8.5 cause variation in the T and variabilities in the parameters marked in the portion 'D' in Figure 8.5 cause variation in the N. Therefore, T and N can be represented in the form of a distribution, as shown in Figure 8.6. The distributions of T and N can be obtained analytically or numerically by simulation. One can refer to, for example [115, 192, 284], for more details.

It can be said that if N always assumes higher value than T, then the pavement will be safe during the design period; hence reliability will be 100%. If, however, there is some overlap, (as is shown in Figure 8.6), the reliability value will be lower than 100%. Recalling Equation 8.1, a term damage ratio, D, can be defined as

$$D = \frac{T}{N} \tag{8.5}$$

The reliability (R) can be defined as

$$\begin{aligned} R &= \text{Probability}(T < N) \\ &= \text{Probability}(D < 1) \end{aligned} \tag{8.6}$$

The reliability value can be estimated from the distribution of D (refer Figure 8.6). The distribution of D can be derived from known distributions of T and N analytically [61, 164, 236] or by simulation [67, 192, 285].

For a pavement design problem, the position of T is rather fixed, a designer can move position of N (that is by iteratively varying the thickness) so that the reliability

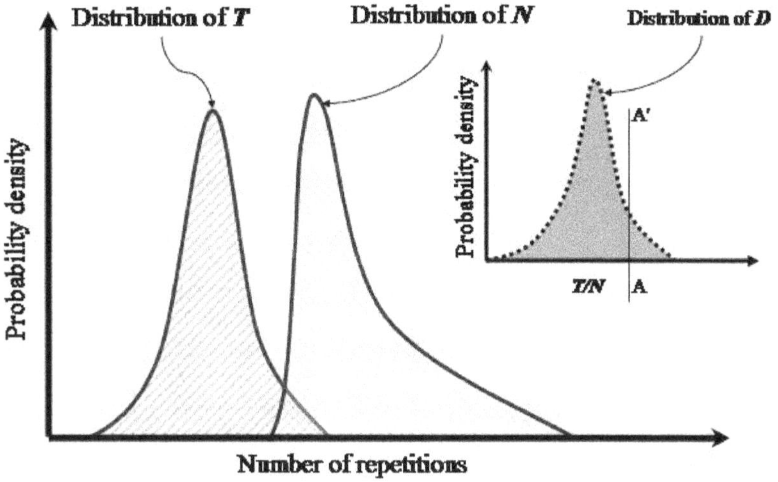

Figure 8.6 Estimation of reliability from distributions of N and T

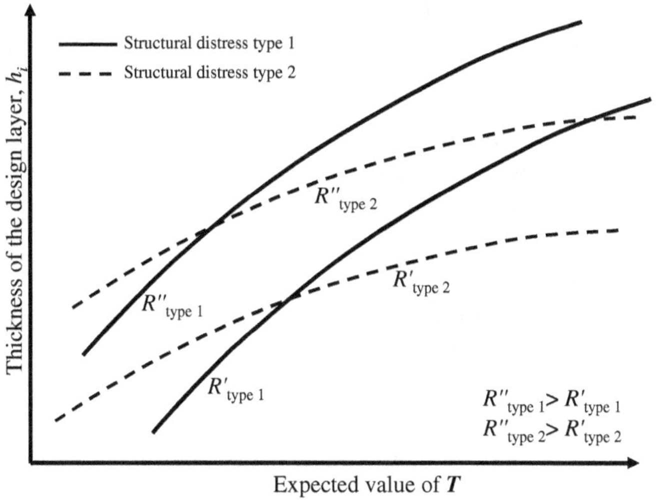

Figure 8.7 Schematic diagram of pavement design chart

attains the designed reliability level. One can refer to [237] for a suggested procedure for designing a pavement for a given reliability level. It can be said that if the design reliability is increased, larger thickness will be required for any given pavement layer (h_i) for a given structural distress type. This is schematically shown in Figure 8.7.

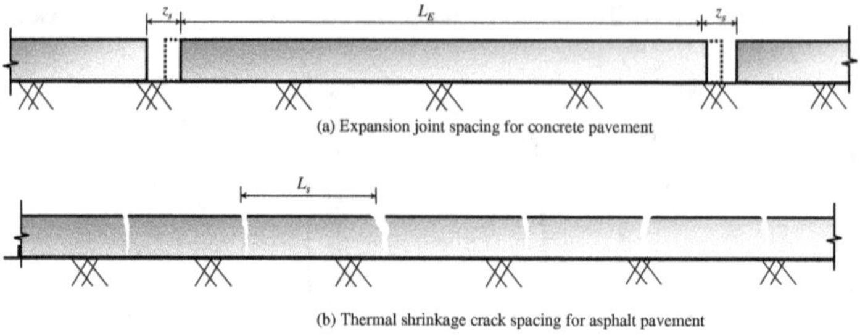

Figure 8.8 Schematic diagram explaining the thermal expansion and contraction joint spacing

8.4.3 DESIGN OF JOINTS

Design of joint involves considerations of joint spacing, design of dowel and tie bars. These are discussed in the following.

8.4.3.1 Estimation of Joint Spacing

In concrete pavement, expansion joints are provided to allow expansion due to temperature variations.

Joint spacing for expansion joint

If the concrete pavement is freely allowed to expand or contract horizontally (refer Figure 8.8(a)), from Equation 4.10 one can write

$$L_E \alpha \left(T^A - T_o\right) = z_s \tag{8.7}$$

where L_E is the spacing between the two successive expansion joints (which eventually is the length of the concrete slab). If the concrete pavement is partially restrained to move, then from Equation 4.33 one can write

$$\alpha \left(T^A - T_o\right) \frac{e^{-\xi L_E}}{\xi(1 + e^{-\xi L_E})} = z_s \tag{8.8}$$

The above equations provides a relationship between z_s and L_E. Thus, the spacing of expansion joints, L_E, is obtained by setting limits to its displacement in terms of joint gap, z_s. If z_s provided is too small concrete slabs may buckle at the joints (known as blow up), if it is too large it may cause hindrance to the smooth movement of wheels while moving from one slab to the other (refer Figure 3.6).

Joint spacing for thermal shrinkage crack

As discussed earlier, the thermal shrinkage crack (refer Figure 8.8(b)) happens at that instant of time when the tensile thermal stress generated exceeds the tensile strength of the material [83]. Thus, the spacing of the contraction joint, L_C, can be obtained by setting limit to the tensile stress developed as the tensile strength. That is,

$$\sigma^T(z,t) = \sigma^S \tag{8.9}$$

where σ^S is the tensile strength of the asphalt mix. Thus, assuming a full-depth thermal shrinkage crack (refer Figure 8.8(b)), Equation 4.17 can be used to obtain the value of L_C, if a friction model is used [198, 235, 264, 343].

Researchers have found that the σ^S of asphalt mix is sensitive to temperature [122, 312]. Approaching from low temperature (say, 40°C) to higher temperature, the tensile strength of asphalt material is first observed to increase then decrease [312].

The above principle also holds for estimation of crack spacing due to natural shrinkage of concrete (known as contraction joint). However, stress needs to be calculated differently than the way done in Equations 4.11 and 4.33 due to thermal variations. The restrained strain is due to shrinkage potential (which in principle may be independent of temperature) and that would give rise to stress.

Contraction joints can be provided in variety of ways [328]. Primarily contraction joint is built as transversely placed pre-cracks (at regular intervals as estimated above), so that if shrinkage crack grows further, they will occur along the same transverse lines. This will prevent irregular growth of crack on the pavement surface (also refer to the discussions in Section 1.1).

8.4.3.2 Design of Dowel Bar

The dowel bars (refer Figure 1.3) participate in the load transfer. For the design of dowel bar, the portion of the load transmitted by a single dowel bar is to be estimated first. This can be done by assuming that the dowel bar which is placed below the wheel takes the maximum share of the wheel load, and it is assumed that the subsequent dowel bars (up to some assumed length) take share load following a similar triangular rule [128, 328]. It may be noted that the total load shared by all the participating dowel bar is a fraction of the wheel load acting on the slab – the remaining load gets transmitted through the slab to the underlying layer [328].

Having obtained the maximum possible load on a single dowel bar (that is, load Q in Figure 3.6), the dowel bar can be analysed (refer Section 3.2 for the principles) for bending, bearing and shear, and accordingly its cross section (for a given type of steel) can be decided. The design can be finalized by varying the diameter or spacing or both.

8.4.3.3 Design of Tie Bar

The tie bars (refer Figure 1.3) are used for keeping two adjacent slabs tied to eachother. Figure 8.9 schematically shows a single tie bar.

The strength of the tie bar should be just adequate so that it does not snap when the slab is pulled (such a situation may arise due to any lateral displacement of a slab). Considering unit length of the slab, one can write

$$f\rho g(B \times h \times 1) = \sigma_{st}.A_{st} \tag{8.10}$$

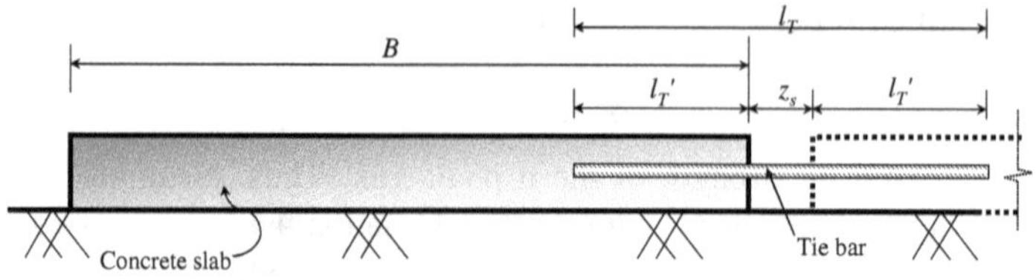

Figure 8.9 Schematic diagram of a single tie bar

where f = coefficient of friction between the concrete slab and the underlying layer, ρ = density of concrete material, g = gravitational acceleration, h = thickness of the concrete slab, σ_{st} = tensile strength of steel, A_{st} = area of the steel per unit length. Equation 8.10 provides the total area of steel required per unit length of the concrete slab. Further, considering pull-out strength of a single tie bar, one can write

$$a_{st}\sigma_{st} = l'_T.2\pi a.\tau^b \tag{8.11}$$

where a_{st} = cross sectional area of a single tie bar, l'_T is the length embedded inside one of the slabs (refer Figure 8.9)), a = radius of the tie bar, τ^b = bond strength between tie bar and concrete. Thus, the total length of the tie bar can be calculated as

$$l_T = 2l'_T + z_s \tag{8.12}$$

where z_s = gap between the two adjacent slabs.

8.5 LIFE CYCLE COST ANALYSIS AND REHABILITATION STRATEGY

A pavement which is designed for a specified design period will undergo deterioration. There is a need to maintain the pavement time to time. This maintenance strategy may involve minor maintenance, repair or rehabilitation. It may involve recycling or may even require reconstruction. The choice on the type of maintenance and its timing is an interesting problem [298]. This is briefly introduced in the following,

Figure 8.10 shows schematic diagram of variation of structural health of pavement over maintenance cycles. The beginning of each cycle may be considered to start with a maintenance activity.

Referring to a typical ith cycle (between time t^i and t^{i+1}) shown in Figure 8.10, the structural health is improved by ΔS^i (that is, structural health is improved from S^i_S to S^i_E) due to maintenance. The cost for the maintenance (that is, Agency cost) may be considered as

$$\text{Agency cost} = \Delta S^i M_c \quad \text{for } t^i < t \leq t^{i+1} \tag{8.13}$$

where M_c is the cost of maintenance for unit improvement of the structural health of the pavement. After each maintenance, the structural health will drop gradually (to

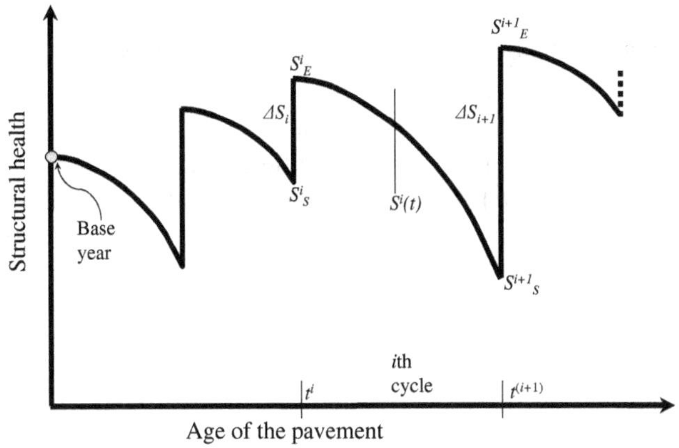

Figure 8.10 Schematic diagram showing variation of structural health of pavement over maintenance cycles

S_S^{i+1}, in this case) until the next maintenance is initiated and so on. One may consider that this trend may be given by a function as

$$S^i(t) = S^i_E f(t) \quad \text{for } t^i < t \leq t^{i+1} \tag{8.14}$$

One may further assume that the deterioration of road condition causes increase in road users' cost (in terms of increase in wear and tear of tyre, increase in the consumption of fuel due to road roughness, increase in the travel time and so on). Thus, the road users' cost can be calculated as [174]

$$\int_{t^i}^{t^{i+1}} S^i_E f(t') M_u dt' \quad \text{for } t^i < t \leq t^{i+1} \tag{8.15}$$

where M_u is road users' cost. Now, one may add the agency cost and user cost to obtain the total cost, and further add these costs for all the cycles. Since these costs will be incurred in different points of time, one needs to consider discounted cost at a specific base year. If the year in which the pavement is opened to the traffic is considered as the base year (refer Figure 8.10), then one can write the total cost as [174]

$$\text{TDC} = \sum_{\forall i} \left(\Delta S^i M_c e^{-rt^i} + \int_{t^i}^{t^{i+1}} S^i_E f(t') M_u e^{-rt'} dt' \right) \tag{8.16}$$

where TDC is the total discounted cost, r is the discount rate.

For a given road project, one may like to minimize the total discounted cost and obtain the optimal maintenance strategy and timing for a steady-state situation [174, 206]. However, there are complexities associated with the real-life problems, for example,

- There are multiple pavement stretches in a road network and one may like to develop an optimal maintenance scheme for the entire network [165, 219], given for each of the pavement stretches,
    - the health status may be different.
    - the deterioration trends may be different and may not be deterministically known [57, 125]. There is an uncertainty of the predicted longevity of the pavement due to the inherent variability associated with the pavement design parameters [75, 205, 253].
    - the maintenance needs of individual stretches may be different, and and each stretch may have multiple maintenance alternatives [50].
    - the maintenance priorities may be different [85].
    - and so on.
- There may be constraint in terms of resources. These resource constraint may include budget, manpower and equipment constraint [50].
- The pavement health parameters for taking a decision choice of maintenance options [134] among preventative maintenance [228], repair, rehabilitation, recycling, reconstruction need to be identified.
- One may like to take the energy and green house gas emissions in consideration [18] while deciding from amongst the maintenance alternatives [177].
- and so on.

8.6 CLOSURE

In this chapter, an overview is presented on the principles used in structural design of pavement. Like any other design process pavement design is also an iterative process in which the best thickness combinations are estimated out of several possible design alternatives. The design alternative arises because of choices available in the selection of layer compositions, their relative costs and the reliability level achieved [237]. However, the readily available design charts and softwares provide support to a pavement designer to finalize the pavement design [1, 214, 220, 224, 225, 292, 295].

Pavement design uses the analysis (generally for static load conditions) results to obtain the predicted critical stress–strain values. For most of the distresses, failure of a pavement is a function of traffic repetition (or environmental cycles). And these are generally linked through empirical relationships, and the calibration essentially depends on the local conditions [211, 293].

Given that pavement design is primarily governed by the number of traffic repetitions (and not by ultimate load bearing conditions, except for few types of distresses), it becomes critical to understand how damage propagates due to repetitions (of traffic and environmental cycls) for these distresses. Implementation of such understanding in the analysis process would eventually minimize the uncertainties in the pavement performance.

9 Miscellaneous Topics

9.1 INTRODUCTION

Some miscellaneous analysis are discussed in this chapter. This includes beam resting on half-space, plates/beams resting on elastic foundation and subjected to dynamic loading, analysis of composite pavements and inverse problem in pavement engineering.

9.2 PLATES/BEAMS RESTING ON ELASTIC FOUNDATION SUBJECTED TO DYNAMIC LOADING

On a pavement structure load is applied as a pulse [23]. The governing equation of a horizontal beam (Euler–Bernoulli beam) resting on an elastic foundation (represented by lumped parameter model) subjected to a concentrated oscillating force of $Q\cos w_f$ acting at an angle of β with the horizontal line and moving with a speed V_o (refer Figure 9.1) can be written as

$$\begin{aligned} EI\frac{\partial^4 w}{\partial x^4} + \rho A\frac{\partial^2 w}{\partial t^2} + C_d\frac{\partial^4 w}{\partial x^4} + Q\cos(w_f t).\cos\beta\frac{\partial^2 w}{dx^2} \\ = Q\cos(w_f t).\sin\beta.\delta(x - V_o t) + q* \end{aligned} \quad (9.1)$$

where EI = flexural rigidity of the beam, ρ = density of the beam, A = cross-sectional area, C_d = coefficient of damping δ = Dirac delta function. One can see that when load is constant (that is, $Q\cos(w_f t) = q$), vertical (that is, $\beta = 0$) and stationary (that is, $V_o = 0$), the Equation 9.1 reduces to Equation 3.11. A number of research publications are available where researchers have derived steady-state closed-form solution of such equation; for example, solutions for

- a vertical concentrated oscillating load of moving with constant speed on an infinite Euler–Bernoulli beam resting on Winkler's foundation [202].
- a vertical concentrated load of constant magnitude moving with constant speed and a constant horizontal axial load on an infinite Euler–Bernoulli beam resting on Winkler's foundation [150].
- a vertical concentrated load of constant magnitude moving with constant speed on an infinite Euler–Bernoulli beam resting on Vlasov foundation [194] or a viscoelastic foundation [27, 277].
- a vertical concentrated load of constant magnitude moving with constant speed on an infinite Timoshenko beam resting on viscoelastic foundation [53].
- a tandem-axle with varying amplitude moving on a plate resting on viscous Winkler's foundation [158].

DOI: 10.1201/9781003190769-9

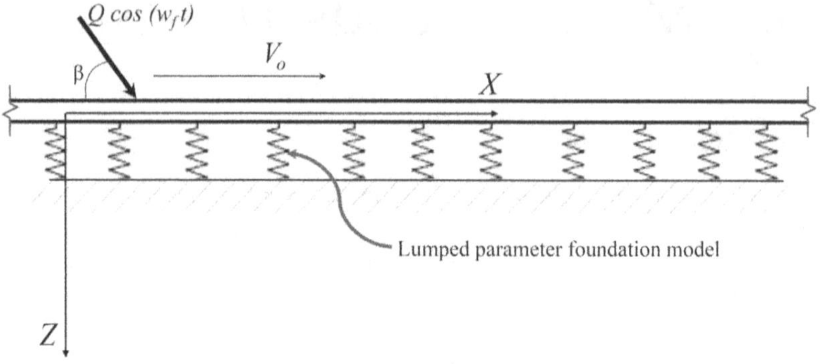

Figure 9.1 An infinite beam subjected to a moving dynamic load

- a point load moving over a half-space [99] or, an oscillatory load over layered over a layered half-space [108] or, an oscillatory load moving over a layered visco-elastic continuum [80].
- and so on.

One may refer to, for example, [31] for a review or [48, 94, 199] for a detailed discussion on this topic.

For a static concentrated vertical oscillating load of $Q\cos(w_f t)$ acting on an infinite beam resting on a Winkler's foundation, the solution is obtained as [202]

$$w = \frac{Q\lambda'}{2(k - A\rho w_f^2)} e^{-\lambda' x} \left(\cos\lambda' x + \sin\lambda' x\right) \cos(w_f t) \tag{9.2}$$

where $\lambda' = \left(\frac{k - A\rho w_f^2}{4EI}\right)^{\frac{1}{4}}$. It is interesting to compare Equation 9.2 with Equation 3.5. From Equation 9.2, it can be shown that amplitude becomes very large (that is, resonance occurs) when

$$w_f^2 = \frac{k}{A\rho} \tag{9.3}$$

One of the ways to obtain a solution for visco-elastic foundation is by using elastic-viscoelastic correspondence principle briefly discussed in Sections 2.3.1.4 and 5.4. One can refer to [88], for example, for an illustrative example on response of an infinite Euler–Bernoulli beam resting on a viscoelastic foundation subjected to a load $Q(x,t)$ along its length.

For a constant load Q moving with constant speed V_o, the deflection is obtained as [202]

$$w = \frac{Q}{2EI} \frac{e^{-c_2(x - V_o t)}}{2c_1 c_2 (c_1^2 + c_2^2)} \left(c_1 \cos c_1 (x - V_o t) + c_2 \sin c_1 (x - V_o t)\right) \tag{9.4}$$

where c_1 and c_2 are constants expressed as functions of k, ρ, A, V_o, E and I [202]. The deflection under the load Q can be calculated by putting $x = 0$ in Equation 9.4, and it can be shown that the deflection due to load Q moving with constant speed V_o is larger than deflection under the static load Q (refer Equation 3.7) [202].

9.3 ANALYSIS OF COMPOSITE PAVEMENTS

Composite pavements are those where, in principle, any combination of asphaltic, cement concrete or cemented layer can be provided anywhere within the pavement structure (as also mentioned in Section 1.1). Thus, the formulation will involve consideration of spring, plate or continuum layer anywhere in the pavement structure. One may refer to [133, 159, 161] for an approach to solve such problems.

For analysis of such a pavement, the individual governing equations remain same. That is, for plate Equation 3.36 and for continuum layer or half-space Equation 5.10 (or Equation 5.16 depending on the choice of coordinate) need to be used for concrete layer and bituminous/granular layer, respectively. If an ith layer consists of a set of springs, then it will be governed by (similar to Equation 2.6),

$$\sigma_{zz}^{i} = k\left(w^{(i-1),b} - w^{(i+1),t}\right) \tag{9.5}$$

Equation 5.38 can also be assumed as the ϕ function for the solution [133, 159]. In line with the deflection profile of the multi-layered structure (refer Equation 5.46).

The solution can be obtained in the same manner as discussed in Section 5.4, and the boundary conditions can be imposed after Henkel transformation [133].

For the top surface, the same boundary conditions as mentioned in Equation 5.40 hold. If the top surface is a plate, it may be noted that $\sigma_{zz}^{1,t}$ is the same parameter used as q in the expression of $q^* = q - p$ in the plate equation (that is, Equation 3.36).

The interface conditions mentioned in Section 5.4 can be used. When additional boundary conditions arise, for example, for an interface between a continuum layer (say, ith layer) and a plate (say, $(i+1)$th layer), the conditions can be written as [133, 159, 161]

$$\begin{aligned} \tau_{rz}^{i,b} &= 0 \\ \tau_{rz}^{(i+1),t} &= 0 \\ w^{i,b} = w^{(i+1),t} &= w^{(i+1),b} \end{aligned} \tag{9.6}$$

This is because, it can be assumed that shear stresses are dissipated at the boundary between the plate and the continuum layer, and the vertical deformations at the interface is equal (it also remains the same throughout the plate). If the $(i+1)$th layer is a spring instead of plate, then the first two conditions above may be assumed to hold, and, instead of the third condition, one can write [133]

$$\sigma_{zz}^{i} = \sigma_{zz}^{(i+1)} \tag{9.7}$$

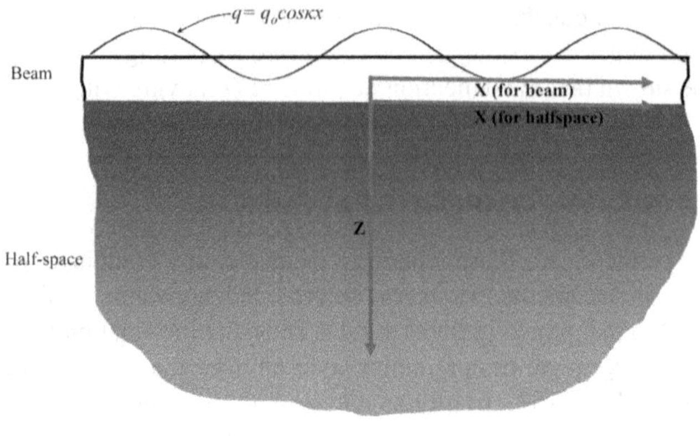

Figure 9.2 An infinite beam resting on elastic half-space

If two consecutive layers are plates, then these can be replaced by one equivalent plate, depending on whether interface is smooth or rough [45, 128, 159]. If two consecutive layers are springs, then these two can be replaced with a equivalent spring [133].

9.3.1 PLATE/BEAM RESTING ON HALF-SPACE

A plate or a beam resting on a half-space is a special case for the generalized formulation for the composite pavement discussed in the above. The geometry of the problem may consist of an infinite beam resting on a half-space (plane stress problem) or circular plate resting on a half-space (axi-symmetric problem).

Biot [32] provided a solution for an infinite beam resting on a half-space – and this is discussed here. Figure 9.2 shows an infinite beam (of unit width) resting on a half-space acted upon a sinusoidal loading as

$$q = q_o \cos \kappa x \tag{9.8}$$

It is assumed [32] that this loading also produces a sinusoidal pressure distribution on the half-space, given as

$$p = p_o \cos \kappa x \tag{9.9}$$

The analysis can be done in two parts, half-space analysis (refer Chapter 5) and beam on elastic foundation analysis (refer Chapter3). For the elastic half-space, the ϕ function (refer Equation 5.10) is assumed[1] as [32]

$$\phi = \frac{p_o}{\kappa^2} \cos \kappa x e^{-\kappa z} (1 + \kappa z) \tag{9.10}$$

[1]by initially assuming $\phi = \cos(\kappa x) F(z)$, where $F(z) = Ae^{\kappa z} + Be^{-\kappa z} + Cze^{\kappa z} + Dze^{-\kappa z}$, and then using the boundary conditions on stresses.

The stresses can be calculated using Equation 5.9. The boundary conditions are $\sigma_{xx} = \sigma_{zz} = \tau_{xz} = 0$ for $z \to \infty$ and $\sigma_{zz} = -p_o \cos \kappa x$. The σ_{xx} and σ_{zz} are obatined as [32]

$$\sigma_{xx} = p_o\,(1 - \kappa z)\,e^{-\kappa z}\cos \kappa x$$
$$\sigma_{zz} = p_o\,(1 + \kappa z)\,e^{-\kappa z}\cos \kappa x$$

Thus [32],

$$\begin{aligned} w = \int_0^\infty \varepsilon_z dz &= \frac{1}{E}\left(\sigma_{zz} - \nu\sigma_{xx}\right) \\ &= \frac{2p_o}{E\kappa}\cos \kappa x \end{aligned} \tag{9.11}$$

where E is the elastic modulus of the half-space. Putting the value of w to Equation 9.9, one obtains

$$p = \frac{1}{2}E\kappa w \tag{9.12}$$

The equation of the beam will be the same as Equation 3.11. It may be noted that the deflection of the beam will be the same as the deflection of the surface of the halfspace. The beam does not have any variation of deflection across its depth (in the current beam model used). Putting Equation 9.8 and Equation 9.12 in Equation 3.11, one obtains

$$E_b I \frac{d^4 w}{dx^4} = q_o \cos \kappa x - \frac{1}{2}E\kappa w \tag{9.13}$$

where E_b is the elastic modulus of the beam[2]. The solution of the above equation is obtained as [32],

$$w = \frac{q_o \cos \kappa x}{2E_b\kappa\left(\frac{1}{2} + \frac{E_b I}{E}\kappa^3\right)} \tag{9.14}$$

Thus, the deflection of a beam resting on an elastic half-space due to sinusoidal loading is obtained. One can employ suitable transformation to obtain the response due to other loading and can refer to [32, 243, 258] for further reading, including alternative methods of solving these set of problems.

A number of studies are available on analysis of plates resting half-space. The deflection of a plate resting on a half-space acted upon by load Q uniformly distributed over a circular area of radius a is given as [123, 229, 286, 325],

$$w = \frac{2(1-\nu^2)}{E\pi a}\int_0^\infty \frac{Q}{m(1+m^3 l_o^3)} J_1(ma) J_0(mr)\,dm \tag{9.15}$$

where $l_o^3 = \frac{2D(1-\nu^2)}{E}$, and $D =$ flexural rigidity of plate. Interested readers can refer to, for example, [117] for a treatise on this topic.

The problem of a beam or a plate on a halfspace or continuum layer has also been solved using minimization of potential energy [141, 218, 304, 313].

---

[2]To differentiate it from the elastic modulus of soil, E.

Dense liquid (that is, springs or lumped parameter models) and continuum (that is, half-space) are two alternative models (characterized by k and E respectively) used to represent soil subgrade. Conventionally, k is used for analysis of concrete pavements (refer Chapters 3 and 4) and E is used for analysis of asphalt pavements (refer Chapters 5 and 6), because of mathematical compatibility with the respective equations. Researchers have opined that the actual field behaviour possibly lies somewhere in between the responses predicted by these two models [96, 130]. It is argued that the k value at a point is dependent only on the vertical displacement at this point, whereas for the continuum model is dependent on the wavelength (refer Equation 9.12), therefore involves interaction from the other parts of the medium too [32, 159]. It is interesting to examine an equivalency between E and k that may exist.

For a perfectly rigid plate of radius a acted upon by a uniform pressure q, one can write

$$w = \frac{q}{k} \tag{9.16}$$

If the rigid plate rests on an elastic half-space (of elastic modulus E), where a force Q causes an average pressure[3] as q_{av} (that is $q_{av} = \frac{Q}{\pi a^2}$), the deflection can be calculated using Equation 5.35. Equating Equations 9.16 and 5.35, one obtains [260]

$$k = \frac{2E}{\pi(1-\nu^2)a} \tag{9.17}$$

One may refer to [260] for further discussions on this topic.

9.4 INVERSE PROBLEM IN PAVEMENT ENGINEERING

An analysis provides information on the expected response of the pavement. The pavement may be subjected to physical load [334], stress waves [84], electromagnetic wave (infrared wave, ground penetrating radar [44]), etc. For an idealized multi-layered structure (with known geometry and material properties), theories have been formulated to predict the response due to static (as discussed in Sections 5.4 and 3.4) and dynamic load (refer Section 9.2) and so on.

During the structural evaluation of pavements, reverse situation arises. For example, the pavement may be subjected to an impact load and the information on its response (in terms of the instantaneous surface deflection) is collected, and the objective is to predict the parameters of the pavement structure. This parameters may include layer thicknesses, physical (for example, density, thermal conductivity, etc.) and mechanical (for example, material constants needed to describe the constitutive behaviour) properties of the individual layers. This is an example of an inverse problem in pavements. The response typically contains a mixed-up up information of all the layers; so, the task is to decipher the individual properties of the each layer.

Typically, an inverse problem (also known as, back-calculation) algorithm involves (i) assuming of initial values for the unknown parameters and performing

[3] it may be noted that the pressure distribution is expected to be non-uniform as discussed in Section 5.3.

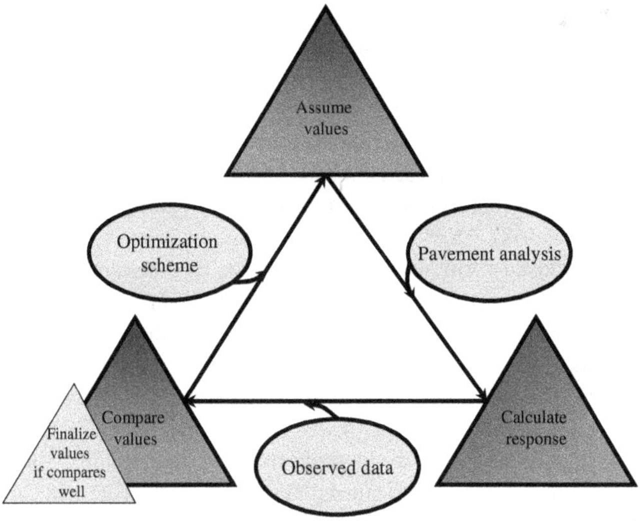

Figure 9.3 Schematic diagram explaining an approach to solve inverse problem in structural evaluation of pavements

analysis (that is, forward calculation), (ii) comparing the analysis results with the field observations and (iii) revising the trial values of the unknown parameters for the next iteration, until the observed and calculated analyses results become comparable (in terms of any suitable error function), and subsequently convergence is said to have achieved. Figure 9.3 provides a schematic diagram explaining this approach.

Various optimization (classical and evolutionary) schemes are suggested to perform iteration and subsequent finalization of unknown parameters. The complexity (in terms of problem formulation, convergence, computational time, etc.) increases as the number of unknowns increases. For example, it is more difficult to back-calculate the parameters for the non-linear constitutive models of the individual pavement layers of a n layered structure, than obtaining, (say) the elastic moduli of a pavement structure made with two homogeneous isotropic continuum layers. Clearly, inverse solutions are quite straight-forward when closed-form analysis results are available (for example, refer to Equations 3.5, 5.32, etc) and may not involve any iteration.

With an assumed pavement structure with known material properties and an associated theory, it is generally possible to uniquely obtain the predicted response within the framework of idealizations involved; however, reverse is not true. Thus, solution to the inverse problem may give rise to non-uniqueness [95], that is, multiple answers are possible within the permissible error-band.

9.5 CLOSURE OF THE BOOK

Basic formulations for estimation of load and thermal stress in concrete and asphalt pavements are dealt in this book. The analyses involved a number of idealizations.

Most of the time it has been assumed that the pavement materials behave like a linear, homogeneous elastic, structure, the interface conditions are either perfectly smooth or rough (fully bonded), the load is static, that the life of the pavement can be predicted from initial stress–strain conditions, relative damage is linearly accumulative and so on. However, pavement materials behave in a more complex way than what are generally considered in idealized analysis schemes and propagation of pavement distresses are more complex a phenomena than what are generally considered in routine design processes [68].

Obtaining a closed-form solution for analysis of a pavement structure is not always quite easy because of involvement of multiple layers in the formulation. The difficulty level increases further, when simplified assumptions on geometry, loading and material properties are replaced with more realistic ones, or, a new material is proposed to be used in pavement construction. These can be handled by invoking various numerical methods, as FEM as discussed in Chapter 7. Efforts are being made to develop comprehensive multi-scale model of pavement which can simulate deformation, damage and its location, healing and so on [8, 100, 200]. Further, it is important to validate the analysis results from field studies through appropriate instrumentation. Significant initiatives have also been undertaken in this direction [12, 28, 81, 129, 180, 262, 327].

The gap between the theoretical and experimental results is gradually reducing through advanced modelling and instrumentation [13, 248]. This would eventually bring down the level of uncertainty. Researchers are trying to understand the pavements better.

Bibliography

1. *AASHTO Guide for design of pavement structures*, American Association of State Highway and Transportation Officials, Washington, D. C., 1993.
2. *AASHTO-T321, Standard method of test for determination the fatigue life of compacted asphalt mixtures subjected to repeated flexural bending*, AASHTO, Washington, D. C., 2017.
3. *AASHTO T307-99, Standard method for test for determining the resilient modulus of soils and aggregate materials*, 2007, Washington, D. C.
4. *AASHTO TP 79-10, Standard method of test for determining dynamic modulus and flow number for hot mix asphalt using the asphalt mixture performance tester (AMPT)*, AASHTO, Washington, D.C., 2010.
5. ACI Committee 318, *Building code requirements for structural concrete and commentary*, American Concrete Institute, 2008.
6. Acum, W. E. A., and Fox, L., Computation of load stresses in a three-layered elastic system, *Géotechnique*, 2(4), 1951, pp. 293–300.
7. Alanazi, N., Kassem, E., Grasley, Z., and Bayomy, F., Evaluation of viscoelastic Poisson's ratio of asphalt mixtures, *International Journal of Pavement Engineering*, 20(10), 2019, pp. 1231–1238.
8. Allen, D. H., Little, D. N., Soares, R. F., and Berthelot, C., Multi-scale computational model for design of flexible pavement – part I: expanding multi-scaling, *International Journal of Pavement Engineering*, 18(4), 2017, pp. 309–320.
9. Al-Rub, R. K. A., Darabi, M. K., Huang, C.-W., Masad, E. A., and Little, D. N., Comparing finite element and constitutive modelling techniques for predicting rutting of asphalt pavements, *International Journal of Pavement Engineering*, 13(4), 2012, pp. 322–338,
10. Al-Qadi, I.L., Hassan, M. M., and Elseifi, M. A., Field and theoretical evaluation of thermal fatigue cracking in flexible pavements, *Transportation Research Record*, 1919, 2005, pp. 87–95.
11. Al-Qadi, I. L., and Nassar, W.N., Fatigue shift factors to predict HMA performance, *International Journal of Pavement Engineering*, 4(2), 2003, pp. 69–76.
12. Anupam, K., Tang, T., Kasbergen, C., Scarpas, A., and Erken, S., 3-D Thermomechanical tire–pavement interaction model for evaluation of pavement skid resistance, *Transportation Research Record*, 2675(3), 2020, pp. 65–80
13. Apeagyei A., Elwardnany M., Hall K., McCarthy L., Timm D., and Quintus H.V., *Design and rehabilitation of asphalt pavements: history and future*, Centennial Paper, TRB, (2019), http://onlinepubs.trb.org/onlinepubs/centennial/papers/AFD60-Final.pdf, last accessed October 13, 2021.

14. Aragão, F., Kim, Y., Lee, J., and Allen, D., Micromechanical model for heterogeneous asphalt concrete mixtures subjected to fracture failure, *Journal of Materials in Civil Engineering*, 23, Special issue: Multiscale and Micromechanical Modeling of Asphalt Mixes, 2011, pp. 30–38.

15. Armedakàs, A. E., *Advanced mechanics of materials and advanced elasticity*, CRC Press, Taylor & Francis Group, 2006.

16. Asbahan, R. E., and Vandenbossche, J. M., Effects of temperature and moisture gradients on slab deformation for jointed plain concrete pavements, *Journal of Transportation Engineering*, 137(8), 2011, pp. 563–570.

17. *Asphalt overlays for highway and street rehabilitation*, Manual Series No. 17, Asphalt Institute, 1983.

18. Aurangzeb, Q., Al-Qadi, I. L., Ozer, H., and Yang, R., Hybrid life cycle assessment for asphalt mixtures with high RAP content, Resources, *Conservation and Recycling*, 83, 2014, pp. 77–86.

19. Baburamani, P., *Asphalt fatigue life prediction models - a literature review*, Australian Road Research Bureau, Report no. 334.

20. Baaj, H., Mikhailenko, P., Almutairi, H., Benedetto, and Di, H., Recovery of asphalt mixture stiffness during fatigue loading rest periods, *Construction and Building Materials*, 158, 2018, pp. 591–600.

21. Balbo, J. T., and Severi, A. A., Thermal gradients in concrete pavements in tropical environment: experimental appraisal, *Transportation Research Record*, 1809, 2002, pp. 12–22.

22. Bandyopadhyaya, R., Das, A., and Basu, S., Numerical simulation of mechanical behaviour of asphalt mix, *Construction and Building Materials*, 22(6), 2008, pp. 1051–1058.

23. Barksdale, R. G., Compressive stress pulse times in flexible pavements for use in dynamic testing, *Highway Research Record*, 345, 1971, pp. 32–44.

24. Bari, J., and Witczak, M. W., Development of a new revised version of the Witczak E* predictive model for hot mix asphalt mixtures, *Proceedings of the Association of Asphalt Paving Technologists*, 75, 2006, pp. 381–423.

25. Bathe, K-J, *Finite element procedures*, Printice-Hall, 1996.

26. Bhasin, A., Branco, V. T. C., Masad, E., and Little, D. N., Quantitative comparison of energy methods to characterize fatigue in asphalt materials, *Journal of Materials in Civil Engineering*, 21(2), 2009, pp. 83–92.

27. Basu, D., and Rao, N. S. V. K., Analytical solutions for Euler-Bernoulli beam on visco-elastic foundation subjected to moving load, *International Journal for Numerical and Analytical Methods in Geomechanics*, 37, 2013, pp. 945–960.

28. Bayat, A., Knight, M. A., and Soleymani, H. R., Field monitoring and comparison of thermal- and load-induced strains in asphalt pavement, *International Journal of Pavement Engineering*, 13(6), 2012, pp. 508–514.

29. Behnke, R., Wollny, I., Hartung, F., and Kaliske, M., Thermo-mechanical finite element prediction of the structural long-term response of asphalt pavements subjected to periodic traffic load: tire-pavement interaction and rutting, *Computers and Structures*, 218, 2019, pp. 9–31.

30. Benedetto, D. H., and Roche, de la C, *State of the art of stiffness modulus and fatigue of bituminous mixtures*, RILEM Report 17, Bituminous binders and mixes, edited by Francken, L., pp. 137–180.

31. Beskou, N. D., and Theodorakopoulos, D. D., Dynamic effects of moving loads on road pavements : a review, *Soil Dynamics and Earthquake Engineering*, 31(4), 2011, pp. 547–567.

32. Biot, M. A., Bending of an infinite beam on an elastic foundation, *Journal of Applied Mechanics*, 1937, pp. A-1-A-7.

33. Bordelon, A., Roesler, J., and Hiller, J., *Mechanistic-empirical design concepts for jointed plain concrete pavements in Illinois*, Research Report ICT-09-052, Illinois Center for Transportation, 2009.

34. Boyce, H. R., A non-linear model for the elastic behaviour of granular materials under repeated loading, *Proceedings of International Conference on Soils under Cyclic and Transient Loading*, Swensea, 1980, pp. 285–294.

35. Bradbury, R. D., Design of joints in concrete pavements, *Proceedings of Highway Research Board*, 12, 1933, pp. 105–141.

36. Bradbury, R. D., *Reinforced concrete pavement*, Wire Reinforcement Institute, Washington, D. C., 1938, pp. 34–41.

37. Brown, S. F., and Pappin, J. W., Analysis of pavement with granular bases, *Transportation Research Record*, 810, 1981, pp. 17–22.

38. Brown, S. F., and Pell, P. S., An experimental investigation of the stresses, strains and deflections in a layered pavement structure subjected to dynamic loads, *Proceedings of 2nd International Conference of Structural Design of Asphalt Pavements*, Ann Arbor, 1967, pp. 487–504.

39. Brown, S. F., and Pell, P. S., A fundamental structural design procedure for flexible pavements, *Proceedings of 3rd International Conference of Structural Design of Asphalt Pavements*, London, Vol. I, 1972, pp. 369–381.

40. Burmister, D. M., The theory of stress and displacements in layered systems and applications to the design of airport runways, *Highway Research Record*, 23, 1943, pp. 126–144.

41. Burmister, D. M., Stresses and displacements in elastic layered Systems, *Proceedings of 2nd International Conference on Structural Design of Asphalt Pavements*, University of Michigan, Ann Arbor, 1945, pp. 277–290.

42. Burmister, D. M., The general theory of stresses and displacements in layered soil systems, I, II, and III. *Journal of Applied Physics*, 16. 1945, pp. 84–94(I), 126–127(II), 296–302(III).

43. Cai, Y., Sangghaleh, A., and Pan, E., Effect of anisotropic base/ interlayer on the mechanistic responses of layered pavements, *Computers and Geotechnics*, 65, 2015, pp. 250–257.

44. Carcione, J. M., Ground-penetrating radar: wave theory and numerical simulation in lossy anisotropic media, *Geophysics*, 61(6), 1996, pp. 1664–1677.

45. Cauwelaert, F. V., *Pavement design and evaluation: the required mathematics and its applications*, Editor: Stet, M., Federation of Belgian Cement Industry, http://pavers.nl/pdf/The%20Required%20Mathematics.pdf.

46. Cauwelaert, F. V., and Eckmann, B., Indirect tensile test applied to anisotropic materials, *Materials and Structures*, 27, 1994, pp. 54–60.

47. Cauwelaert, V. F., Stet, M., and Jasienski, A., The general solution for a slab subjected to centre and edge loads and resting on a Kerr foundation, *International Journal of Pavement Engineering*, 3(1), 2002, pp. 1–18.

48. Cebon, D., *Handbook of vehicle-road interaction*, Swets & Zeitlinger, B. V., reprinted 2000.

49. Ceylan, H., Schwartz, C. W., Kim, S., and Gopalakrishnan, K., Accuracy of predictive models for dynamic modulus of hot-mix asphalt, *Journal of Materials in Civil Engineering*, 21(6), 2009, pp. 286–293.

50. Chakroborty, P., Agarwal, P. K., and Das, A., Comprehensive pavement maintenance strategies for road network through optimal allocation of resources, *Transportation Planning and Technology*, 35(3), 2012, pp. 317–339.

51. Chiasson, A., Yavuzturk, C., and Ksaibati, K., Linearized approach for predicting thermal stresses in asphalt pavements due to environmental conditions, *Journal of Materials in Civil Engineering*, 20(2), 2008, pp. 118–127.

52. Chen, G., and Baker, G., Analytical model for predication of crack spacing due to shrinkage in concrete pavements, *Journal of Structural Engineering*, 130(10), 2004, pp. 1529–1533.

53. Chen, Y. H., Huang, Y. H., and Shih, C. T., Response of an infinite Timoshenko beam on a viscoelastic foundation to a harmonic moving load, *Journal of Sound and Vibration*, 241(5), 2001, pp. 809–824.

54. Chen, E. Y. G., Pan, E., and Green, R., Surface loading of a multilayered viscoelastic pavement: semianalytical solution, *Journal of Engineering Mechanics*, 135(6), 2009, pp. 517–528.

55. Chen, J-S, Lin, C-H, Stein, E., and Horthan, J., Development of a mechanistic-empirical model to characterize rutting in flexible pavements, *Journal of Transportation Engineering*, 130(4), 2004, pp. 519–525.

56. Chen, S., Wang, D., Du, R., and Feng, D., Elastic multilayered pavement under an elliptical vertical load: analytical solutions and program verification, *Road Materials and Pavement Design*, 20(2), 2019, pp. 297–315.

57. Chootinan, P., Chen, A., Horrocks, M. R. and Bolling, D., A multi-year pavement maintenance program using a stochastic simulation-based genetic algorithm approach, *Transportation Research Part A: Policy and Practice*, 40(7), 2005, pp. 725–743.

58. Choubane, B., and Tia, M., Nonlinear temperature gradient effect on maximum warping stresses in rigid pavements, *Transportation Research Record*, 1370, 1992, pp. 11–19.

59. Choubane, B., and Tia, M., Analysis and verification of thermal gradient effects on concrete pavement, *Journal of Transportation Engineering*, 121(1), 1995, pp. 75–81.

60. Christensen, R. M., *Theory of viscoelasticity an introduction*, 2nd edition, Academic Press, 1982.

61. Chua, K. H., Kiureghian, A. D., and Monismith, C. L., Stochastic model for pavement design, *Journal of Transportation Engineering*, 118(6), 1992, pp. 769–786.

62. Claussen, A. I. M., Edwards, J. M., Sommer, P., and Uge, P., Asphalt pavement design - the Shell method, *Proceedings of 4th International Conference of Structural Design of Asphalt Pavements*, Ann Arbor, Vol.1, 1977, pp. 39–74.

63. Collop, A. C., Scarpas, A., Kasbergen, C., and Bondt, A. de., Development and finite element implementation of stress-dependent Elastoviscoplastic constitutive model with damage for asphalt, *Transportation Research Record*, 1832, 2003, pp. 96–104.

64. Costanzi, M., and Cebon, D., Generalized phenomenological model for the viscoelasticity of idealized asphalts, *Journal of Materials in Civil Engineering*, 26(3), 2014, pp. 399–410.

65. Daloglu, A. T., and Girijia, C. V., Values of *k* for slab on Winkler foundation, *Journal of Geotechnical and Geoenvironmental Engineering*, 126(5), 2000, pp. 463–471.

66. Daniel, J. S., and Kim, Y. R., Development of simplified fatigue test and analysis procedure using a viscoelastic continumm damage model, *Proceedings of the Association of Asphalt Paving Technologists*, 71, 2002, pp. 619–650.

67. Darter, M., Khazanovich, L., Yu, T., and Mallela, J., Reliability analysis of cracking and faulting prediction in the new mechanistic-empirical pavement design procedure, *Transportation Research Record*, 1936, 2005, pp. 150–160.

68. Das, A., Structural design of asphalt pavements: principles and practices in various design guidelines, *Transportation in Developing Economies*, 1, 2015, pp. 25–32.

69. Das, A., and Pandey, B. B., Mechanistic-empirical design of bituminous roads : an Indian perspective, *Journal of Transportation Engineering*, 125(5), 1999, pp. 463–471.

70. Das, B. M., *Advanced soil mechanics*, 3rd edition, Taylor & Francis, 2008.

71. Davis, R. O., and Selvadurai, A. P. S., *Elasticity and geomechanics*, Cambridge University Press, 1996.

72. Davids, W. G., Turkiyyah, G. M., and Mohoney, J., EverFE - rigid pavement three-dimensional finite element analysis tool, *Transportation Research Record*, 1629, 1998, pp. 41–49.

73. De Jong, D. L., Peatz, M. G., and Korswagen, A. R., *Computer program BISER, layered systems under normal and tangential loads*, External Report AMSR.0006.73, Koninklijke/Shell-Laboratorium, Amsterdam, the Netherlands.

74. Deacon, J. A., *Fatigue of asphalt concrete*, Ph.D. thesis, University of California, Berkeley, 1963.

75. Despande, V. P., Damnjanvic, I. D., and Gardoni, P., Reliability-based optimization models for scheduling pavement rehabilitation, *Computer-Aided Civil and Infrastructure Engineering*, 25, 2010, pp. 227–237.

76. Diefenderfer, B. K., Al-Qadi, I. L., and Diefenderfer, S. D., Model to predict pavement temperature profile: development and validation, *Journal of Transportation Engineering*, 132(2), 2006, pp. 162–167.

77. Dinev, D., Analytical solution of beam on elastic foundation by singularity functions, *Engineering Mechanics*, 19(6), 2012, pp. 381–392.

78. Dorman, G. M., The extension to practice fundamental procedure for design of flexible pavements, *Proceedings of 1st International Conference of Structural Design of Asphalt Pavements*, Ann Arbor, Michigan, 1962, pp. 785–793.

79. Dunlap, W. S., *A report on a mathematical model describing the deformation characteristics of granular materials*, Technical Report 1, Project No. 2-8-62-27, Texas Transportation Institute, Texas A & M University, 1963.

80. Elhuni, H., and Basu, D., Dynamic soil structure interaction model for beams on viscoelastic foundations subjected to oscillatory and moving loads, *Computers and Geotechnics*, 115, 2019, 103157.

81. Elseifi, M. A., Al-Qadi, I. L., and Yoo, P. J., Viscoelastic modeling and field validation of flexible pavement, *Journal of Engineering Mechanics*, 132(2), 2006, pp. 172–178.

82. Elseifi, M. A., Baek, J., and Dhakal, N., Review of modelling crack initiation and propagation in flexible pavements using the finite element method, *International Journal of Pavement Engineering*, 19(3), 2018, pp. 251–263.

83. Epps, A., Design and analysis system for thermal cracking in asphalt concrete, *Journal of Transportation Engineering*, 126(4), 2000, pp. 300–307.

84. Ewing, W. M., and Jardetzky, W. S., and Press, F., *Elastic waves in layered media*, McGraw-Hill Book Company, 1957.

85. Farhan, J. and Fwa, T. F., Incorporating priority preferences into pavement maintenance programming, *Journal of Transportation Engineering*, 138(6), 2012, pp. 714–722.

86. Fatemi, A., and Yang, L., Cumulative fatigue damage and life prediction theories: a survey of the state of the art for homogeneous materials, *International Journal of Fatigue*, 20(1), 1998, pp. 9–34.

87. Federal Aviation Administration (FAA), Advisory Circulars, Airport pavement design and evaluation, 2009, http://www.faa.gov/airports/engineering/pavement_design/, last accessed November 14, 2021.

88. Findley, W. N., Lai, J. S., and Onaran, K., *Creep and relaxation of nonlinear viscoelastic materials with an introduction to linear viscoelasticity*, Dover Publications Inc., 1989.

89. Fish, J., and Belytschko, T., *A first course in finite elements*, John Wiley & Sons, Ltd., 2007.

90. Francken, L., and Clauwaert, C., Characterization and structural assessment of bound materials for flexible bound structures, *Proceedings of the 6th International Conference on Structural Design of Asphalt Pavements*, University of Michigan, Ann Arbor, 1987, pp. 130–144.

91. *French design manual for pavement structures*, Guide Technique, LCPC and SETRA, Union Des Synducates, DeLindustrie Routiere, France, 1997.

92. Friberg, B. F., Load and deflection characteristics of dowels in transverse joints of concrete pavements, *Proceedings of the Highways Research Board*, 18, Washington, D.C., 1938, pp. 140–154.

93. Friberg, B. F., Design of dowels in transverse joints of concrete pavements, *Transactions of ASCE*, 105, pp. 1076–1095, 1940.

94. Fryba, L., *Vibrations of solids and structures under moving load*, Thomas Telford Ltd., 1999.

95. Fwa, T. F., and Setiadji, B. H., Evaluation of backcalculation methods for nondestructive determination of concrete pavement properties, *Transportation Research Record*, 1949, 2007, pp. 75–82.

96. Fwa, T. F., and Setiadji, B. H., Backcalculation analysis of rigid pavement properties considering presence of subbase layer, TRB 87th Annual Meeting Compendium of Papers DVD, Paper No. 08-0434, 2008.

97. Fwa, T. F., Shi, X. P., and Tan, S. A., Analysis of concrete pavements by rectangular thick plate method, *Journal of Transportation Engineering*, 122(2), 1996, pp. 146–154.

98. Fwa, T. F., and Tan, S. A., $C-\phi$ characterization model for design of asphalt mixtures and asphalt pavements, *ASTM Special Technical Publication*, 1469, 2006, pp. 113–126.

99. Gakenheimer, D. C., and Miklowitz, J., Transient excitation of an elastic halfspace by a point load traveling on the surface, *Journal of Applied Mechanics*, 36, 1969, pp. 505–15.

100. Gamez, A., Hernandez, J. A., Ozer, H., and Al-Qadi, I. L., Development of domain analysis for determining potential pavement damage, *Journal of Transportation Engineering: Part B Pavements*, 144, 2018, 04018030.

101. Garber, N. J., and Hoel, L. A., *Traffic and highway engineering*, West Publishing Company, 2009.

102. Genin, G., and Cebon, D., Failure mechanism in asphalt concrete, *Road Materials and Pavement Design*, 1(4), 2000, pp. 419–450.

103. Gerber, J., Jenkins, K., and Engelbrecht, F., FEM SEAL-3D: development of 3D finite element chip seal models, *International Journal of Pavement Engineering*, 21(2), 2020, pp. 134–143.

104. Ghuzlan, K. A., and Carpenter, S. H., Energy-derived, damage-based failure criterion for fatigue testing. *Transportation Research Record*, 1723, 2000, pp. 141–149.

105. Gibson, R. E., The analytical method in soil mechanics, *Géotechnique*, 24(2), 1974, pp. 115–140.

106. Gillespie, T. D., Karamihas, S. M., Cebon, D., Sayers, M. W., Nasim, M. A., Hansen, W., and Ehsan, N., *Effects of heavy vehicle characteristics on pavement response and performance*, UMTRl 92-2, The University of Michigan, 1992.

107. Gould, P. L., *Introduction to linear elasticity*, 2nd edition, Springer, 1994.

108. Grundmann, H., Lieb, M., and Trommer, E., The response of a layered half-space to traffic loads moving along its surface, *Archive of Applied Mechanics*, 69, 1999, pp. 55–67.

109. Gui, J., Phelan, P. E., Kaloush, K. E., and Golden, J. S., Impact of pavement thermophysical properties on surface temperatures, *Journal of Materials in Civil Engineering*, 19(8), 2007, pp. 683–690.

110. Haider, S. W., Harichandran, R. S., and Dwaikat, M. B., Closed-form solutions for Bimodal Axle Load Spectra and relative pavement damage estimation, *Journal of Transportation Engineering*, 135(12), 2009, pp. 974–983.

111. Hall, K. T., Darter, M. I., and Kuo, C. M., Improved methods for selection of *k* value for concrete pavement design, *Transportation Research Record*, 1505, 1995, pp. 128–136.

112. Han, Z., and Vanapalli, S. K., State-of-the-Art: prediction of resilient modulus of unsaturated subgrade soils, *International Journal of Geomechanics*, 16(4), 2016, 04015104.

113. Harichandran, R. S., Yeh, M. S., and Baladi, G. Y., MICH-PAVE: a nonlinear finite element program for analysis of pavements. *Transportation Research Record*, 345, 1971, pp. 15–31.

114. Harr, M. E., *Foundations of theoretical soil mechanics*, McGraw-Hill Inc, 1966.

115. Harr, M. E., *Reliability based design in civil engineering*, McGraw-Hill Book Company, 1987.

116. Hausman, M. R., *Engineering principles of ground modification*, McGraw Hill, 1990.

117. Hemsley, J. A., *Elastic analysis of raft foundations*, Thomas Telford, 1998.

118. Hermansson, Å., Mathematical model for calculation of pavement temperatures : comparison of calculated and measured temperatures, *Transportation Research Record*, 1764, 2001, pp. 180–188.

119. Hetényi, M., *Beams on elastic foundation*, The University of Michigan Press, 11th Printing, 1979.

120. Hicks, R. G., and Monismith, C. L., Factors influencing the resilient properties of granular materials, *Highway Research Record*, 345, 1971, pp. 15–31.

121. Hiltunen, D. R., and Roque, R., A mechanistic-based prediction model for thermal cracking of asphalt concrete pavements, *Proceedings of the Association of Asphalt Paving Technologists*, 63, 1994, pp. 81–108.

122. Hoare, T. R., and Hesp, S. A. M., Low temperature fracture testing of asphalt binders, *Transportation Research Record*, 1728, 2000, pp. 36–42.

123. Hogg, A. H. A., Equilibrium of a thin plate, symmetrically loaded, resting on an elastic foundation of infinite depth, *Philosophical Magazine*, Series 7, 25, 1938, pp. 576–582.

124. Hong, A. P., Li, Y. N., and Bažant, Z. P., Theory of crack spacing in concrete pavements, *Journal of Engineering Mechanics*, 123(3), 1997, pp. 267–275.

125. Hong, H. P., and Wang, S. S., Stochastic modeling of pavement performance, *International Journal of Pavement Engineering*, 4(4), 2003, pp. 235–243.

126. Horonjeff, R., Mckelvey, F. X., Sproule, W. J., and Young, S. B., *Planning and design of airports*, 5th edition, McGraw-Hill Book Company, 2010.

127. Horvath, J. S., Modulus of subgrade reaction: new perspective, *Journal of Geotechnical Engineering*, 109(12), 1983, pp. 1591–1596.

128. Huang, Y. H., *Pavement analysis and design*, 2nd edition, Pearson Prentice Hall, 2004.

129. Hugo, F., and Martin, A. L. E., *Significant findings from full-scale accelerated pavement testing*, Synthesis No. 325, NCHRP, TRB, Washington, D. C., 2004.

130. Ioannides, A. M., Concrete pavement analysis: the first eighty years, *International Journal of Pavement Engineering*, 7(4), 2006, pp. 233–249.

131. Ioannides, A. M., and Hammons, M. I., Westergaard-type solution for edge load transfer problem, *Transportation Research Record*, 1525, 1996, pp. 28–34.

132. Ioannides, A. M., and Khazanovich, L., Nonlinear temperature effects on multilayered concrete pavements, *Journal of Transportation Engineering*, 124(2), 1998, pp. 128–136.

133. Ioannides, A. M., and Khazanovich, L., General formulation for multilayered pavement systems, *Journal of Transportation Engineering*, 124(1), 1998, pp. 82–90.

134. Irfan, M., Khurshid, M. B., Bai, Q., Labi, S., and Morin, T. L., Establishing optimal project-level strategies for pavement maintenance and rehabilitation - a framework and case study, *Engineering Optimization*, 44(5), 2012, pp. 565–589.

135. IRC:37-2012, *Guidelines for the design of flexible pavements*, 3rd Revision, Indian Roads Congress, New Delhi, 2012.

136. IRC:58-2011, *Guidelines for the design of plain jointed rigid pavements for highways*, 3rd Revision, Indian Roads Congress, New Delhi, 2011.

137. IRC:81-1997, *Guidelines for strengthening of flexible pavements using Benkelman Beam Deflection technique*, Indian Roads Congress, New Delhi, 1997.

138. IRC:115-2014, *Guidelines for structural evaluation and strengthening of flexible road pavements using Falling Weight Deflectometer (FWD) technique*, Indian Roads Congress, New Delhi, 2014.

139. Jaeger, L. G., *Elementary theory of elastic plates*, Pergamon Press, 1964.

140. Jiang, X., Zeng, C., Gao, X., Liu, Z., and Qiu, Y., 3D FEM analysis of flexible base asphalt pavement structure under non-uniform tyre contact pressure, *International Journal of Pavement Engineering*, 20(9), 2019, pp. 999–1011.

141. Jones, R., and Xenophontos, J., The Vlasov foundation model, *International Journal of Mechanical Sciences*, 19, 1977, pp. 317–323.

142. Johnson, K. L., *Contact mechanics*, Cambridge University Press, 1985.

143. Jumikis, A. R., *The frost penetration problem in highway engineering*, Rutgers University Press, 1955.

144. Jumikis, A. R., *Theoretical soil mechanics*, Van Nostrand Reinhold, New York, 1969.

145. Kassem, E., Grasley, Z. C., and Masad, E., Viscoelastic Poisson's ratio of asphalt mixtures, *International Journal of Geomechanics*, 13(2), 2013, pp. 162–169.

146. Kausel, E., Early history of soil-structure interaction, *Soil Dynamics and Earthquake Engineering*, 30(9), 2010, pp. 822–832.

147. Kenis, W. J., *Predictive design procedure, VESYS users manual — An interim design for flexible pavements using VESYS structural subsystem*, FHWA-RD-77-154, Federal Highway Administration, U.S. Department of Transportation, 1978.

148. Kerr, A. D., Elastic and viscoelastic foundation models, *Journal of Applied Mechanics*, 31, 1964. pp. 491–498.

149. Kerr, A. D., A study of a new foundation model, *Acta Mechanica*, 1965, 2, pp. 135–147.

150. Kerr, A. D., The continuously supported rail subjected to an axial force and a moving load, *International Journal of Mechanical Science*, 1972. 14, pp. 71–78.

151. Kerr, A. D., and Kwak, S. S., The semi-infinite plate on a Winkler base, free along the edge, and subjected to a vertical force, *Archive of Applied Mechanics*, 63, 1993, pp. 210–218.

152. Kim, J., General viscoelastic solutions for multilayered systems subjected to static and moving loads, *Journal of Materials in Civil Engineering*, 2011, 23(7), pp. 1007–1016.

153. Kim, J., Lee, H. S., and Kim, N., Determination of shear and bulk moduli of viscoelastic solids from the indirect tension creep test, *Journal of Engineering Mechanics*, 136(9), 2010, pp. 1067–1075.

154. Kim, Y. R., *Modeling of asphalt concrete*, McGraw-Hill Construction, ASCE Press, 2009.

155. Kim, Y. R., Allen, D. H., and Little, D. N., Damage-induced modelling of asphalt mixtures through computational micromechanics and cohesive zone fracture, *Journal of Materials in Civil Engineering*, 17(5), 2005, pp. 477–484.

156. Kim, Y. R., and Little. D. N., Evaluation of healing in asphalt concrete by means of the theory of nonlinear viscoelasticity, *Transportation Research Record*, 1228, 1989, pp. 198–210.

157. Kim, Y. R., Little, D. N., and Lytton, R. L., Fatigue and healing characterization of asphalt mixtures, *Journal of Materials in Civil Engineering*, 15(1), 2003, pp. 75–83.

158. Kim, S-M., and McCullough, B. F., Dynamic response of plate on viscous Winkler foundation to moving loads of varying amplitude, *Engineering Structures*, 25, 2003, pp. 1179–1188.

159. Khazanovich, L., *Structural analysis of multi-layered concrete pavement systems*, Ph.D. thesis, University of Illinois at Urbana-Champaign, 1994.

160. Khazanovich, L., and Wang, Q., MnLayer high-performance layered elastic analysis program, *Transportation Research Record*, 2037, 2007, pp. 63–75.

161. Khazanovich, L., and Ioannides. A. M., DIPLOMAT: an analysis program for both bituminous and concrete pavements, *Transportation Research Record*, 1482, 1994, pp. 52–60.

162. Krishnan, J. M., and Rajagopal, K. R., Review of the uses and modeling of bitumen from ancient to modern times, *Applied Mechanics Review*, 56(2), 2003, pp. 149–214.

163. Krishnan, J. M., Rajagopal, K. R., Masad, E., and Little, D. N., Thermomechanical framework for the constitutive modeling of asphalt concrete, *International Journal of Geomechanics*, 6(1), 2006, pp. 36–45.

164. Kulkarni, R. B., Rational approach in applying reliability theory to pavement structural design, *Transportation Research Record*, 1449, 1994, pp. 13–17.

165. Labi, S., and Sinha, K. C., Life-cycle evaluation of flexible pavement preventive maintenance, *Journal of Transportation Engineering*, 131(10), 2005, pp. 744–751.

166. Lade, P. V., and Nelson, R. B., Modelling the elastic behahiour of granular materials, *International Journal for Numerical and Analytical Methods in Geomechanics*, 11, 1987, pp. 521–542.

167. Lakes, R. S., *Viscoelastic solids*, CRC Press, 1999.

168. Lakes, R. S., and Wineman, A., On Poisson's ratio in linearly viscoelastic solids, *Journal of Elasticity*, 85, 2006, pp. 45–63.

169. Lancellotta, R., *Geotechnical Engineering*, 2nd English Edition, Taylor & Francis, 2009.

170. Lekarp, F., Isacsson, U., and Dawson, A., State of the art. I : resilient response of unbound granular aggregates, *Journal of Transportation Engineering*, 126(1), 2000, pp. 66–75.

171. Lekarp, F., Isacsson, U., and Dawson, A., State of the art - II: permanent strain response of unbound aggregates, *Journal of Transportation Engineering*, 126(1), 2000, pp. 76–83.

172. Levenberg, E., Analysis of pavement response to subsurface deformations, *Computers and Geotechnics*, 50, 2013, pp. 79–88.

173. Levenberg, E., and Skar, A., Analytic pavement modelling with a fragmented layer, *International Journal of Pavement Engineering*, 2020, https://doi.org/10.1080/10298436.2020.1790559

174. Li, Y., and Madanat, S., A steady-state solution for the optimal pavement resurfacing problem, *Transportation Research Part A: Policy and Practice*, 2002, 36(2), pp. 525–535.

175. Liang, R. Y., and Niu, Y-Z, Temperature and curling stress in concrete pavements: analytical solutions, *Journal of Transportation Engineering*, 1998, 124(1), pp. 91–100.

176. Little, D. N., and Nair, S., *Recommended practice for stabilization of subgrade soils and base materials*, Web-only document No. 144, NCHRP, TRB, Washington, D. C., 2009.

177. Liu, X., Cui, Q., and Schwartz, C., Greenhouse gas emissions of alternative pavement designs: framework development and illustrative application, *Journal of Environmental Management*, 132, 2014, pp. 313–322.

178. Liu, W., and Fwa, T. F., Nine-slab model for jointed concrete pavements, *International Journal of Pavement Engineering*, 8(4), 2006, pp. 277–306.

179. Loulizi, A., Flintsch, G.W., Al-Qadi, I.L., and Mokarem, D., Comparing resilient modulus and dynamic modulus of hot-mix asphalt as material properties for flexible pavement design, *Transportation Research Record*, 1970, 2006, pp. 161–170.

180. Loulizi, A., Al-Qadi, I.L., and Elseifi, M., Difference between in situ flexible pavement measured and calculated stresses and strains, *Journal of Transportation Engineering*, 132(7), 2006, 574–579.

181. Love, A. E. H., *Mathematical theory of elasticity*, 1927, 2nd edition, Oxford University Press, 1906.

182. Lu, Z., Yao, H., Liu, J., and Hu, Z., Experimental evaluation and theoretical analysis of multi-layered road cumulative deformation under dynamic loads, *Road Materials and Pavement Design*, 15(1), 2014, pp. 35–54

183. Lundstrom, R., Benedetto, H. D., and Isacsson, U., Influence of asphalt mixture stiffness on fatigue failure, *Journal of Materials in Civil Engineering*, 16(6), 2004, pp. 516–525.

184. Lundstrom, R., Ekblad, J., Isacsson, U., and Karlsson, R., Fatigue modeling as related to flexible pavement design, *Road Materials and Pavement Design*, 8(2), 2007, pp. 165–205.

185. Mackiewicz, P., Thermal stress analysis of jointed plane in concrete pavements, *Applied Thermal Engineering*, 73, 2014, pp. 1169–1176.

186. Madhav, M. R., and Poorooshasb, H. B., A new model for geosynthetic reinforced soil, *Computers and Geotechnics*, 6, 1988, pp. 277–290.

187. Mahboub, K. C., Liu, Y., and Allen, D. L., Evaluation of temperature responses in concrete pavement, *Journal of Transportation Engineering*, 130(3), 2004, pp. 395–401.

188. Maheshwari, P., Chandra, S., and Basudhar, P. K., Modelling of beams on a geosynthetic-reinforced granular fill-soft soil system subjected to moving loads, *Geosynthetics International*, 11(5), 2004, pp. 369–376.

189. Mahrenholtz, O. H., Beam on viscoelastic foundation: an extension of Winkler's model, *Archive of Applied Mechanics*, 2010, 80, pp. 93–102.

190. Maina, J. W., and Matsui, K., Developing software for elastic analysis of pavement structure responses to vertical and horizontal surface loadings, *Transportation Research Record*, 1896, 2004, pp. 107–118.

191. Maina, J. W., Ozawa, Y., and Matsui, K., Linear elastic analysis of pavement structure under non-circular loading, *Road Materials and Pavement Design*, 13(3), 2012, pp. 403–421.

192. Maji, A., and Das, A., Reliability considerations of bituminous pavement design by Mechanistic-Empirical approach, *International Journal of Pavement Engineering*, 9(1), 2008, pp. 19–31.

193. Malárics, V., and Müller, H. S., Numerical investigations on the deformation of concrete pavements, *Proceedings of 7th RILEM International Conference on Cracking in Pavements*, Delft, 2012, pp. 507–516.

194. Mallick, A. K., Chandra, S., and Singh, A. B., Steady-state response of an elastically supported beam to a moving load, *Journal of Sound and Vibration*, 291, 2006, pp. 1148–1169.

195. Mallick, R. B., and El-Korchi, T., *Pavement engineering - principles and practice*, 2nd edition, CRC Press, Taylor & Francis Group, 2013.

196. Mabrouk, G. M., Elbagalati, O. S., Dessouky, S., Fuentes, L., and Walubita, L. F., 3D-finite element pavement structural model for using with traffic speed deflectometers, *International Journal of Pavement Engineering*, 2021, https://doi.org/10.1080/10298436.2021.1932880

197. Marasteanu, M. O., Li, X., Clyne, T. R., Voller, V. R., Timm, D. H., and Newcomb, D. E., *Low temperature cracking of asphalt concrete pavements*, report no MN/RC - 2004-23, submitted by University of Minnesota to Minnesota Department of Transportation, 2004.

198. Marasteanu, M., Zofka, A., Turos, M., Li, X., Velasquez, R., Li, X., Buttlar, W., Paulino, G., Braham, A., Dave, E., Ojo, J., Bahia, H., Williams, C., Bausano, J., Gallistel, A., and McGraw, J., *Investigation of low temperature cracking in asphalt pavements national pooled fund study*, 776, Rep. No. MN/RC 2007-43, Dept. of Civil Engineering, 2007, University of Minnesota, Minneapolis.

199. Martinček, G., *Dynamics of pavement structures*, E & FN Spon, 1994.

200. Masad, E., Al-Rub, R. A., and Little, D. N., Recent developments and applications of pavement analysis using nonlinear damage (PANDA) model, *Proceedings of 7th RILEM International Conference on Cracking in Pavements*, RILEM Bookseries, Delft, 4, 2012, pp. 399–408.

201. Masad, E., Branco, V. T. F. C., Little, D. N., Lytton, R., A unified method for the analysis of controlled-strain and controlled-stress fatigue testing, *International Journal of Pavement Engineering*, 9(4), 2008, pp. 233–246.

202. Mathews, P. M., Vibrations of a beam on elastic foundation, *Journal of Applied Mathematics and Mechanics*, 38(3-4), 1958, pp. 105–115.

203. Matsuda, H., and Sakiyama, T., Analysis of beams on non-homogeneous elastic foundation, *Computers and Structures*, 25(6), 1987, pp. 941–946.

204. May, R., and Witczak, M. W., Effective granular modulus to model pavement responses, *Transportation Research Record*, 810, 1981, pp. 1–9.

205. McDonald, M., and Madanat, S., Life-cycle cost minimization and sensitivity analysis for mechanistic-Empirical pavement design, *Journal of Transportation Engineering*, 138(6), 2012, pp. 706–713.

206. Meneses, S., and Ferreira, A., Pavement maintenance programming considering two objectives: maintenance costs and user costs, *International Journal of Pavement Engineering*, 14(2), 2013, pp. 206–221.

207. *Mechanistic-empirical pavement design guide, - manual of practice*, AASHTO, Interim Edition, July 2008.

208. Miller, J. S., and Bellinger, W. Y., Distress Identification Manual for Long-Term Pavement Performance Program (Fourth revised edition), FHWA-HRT-13-092, Federal Highway Administration, 2014, https://www.fhwa.dot.gov/publications/research/infrastructure/pavements/ltpp/13092/13092.pdf, last accessed November, 2021.

209. Miner, M. A., Cumulative damage in fatigue, *Proceedings of ASME*, ASME, 1945, pp. A159–A164.

210. Mitchell, J. K., and Monismith, C. L., A thickness design procedure for pavements with cement stabilized bases and thin asphalt surfacings, *Proceedings of 4th International Conference of Structural Design of Asphalt Pavements*, Ann Arbor, Vol. I, 1977, pp. 409–416.

211. Monismith, C. L., Analytically based asphalt pavement design and rehabilitation: theory to practice, 1992-1992, *Transportation Research Record*, 1354, 1994, pp. 5–26.

212. Monismith, C. L., Secor, K. E., and Blackmer, W., Asphalt mixture behaviour in repeated flexure, *Proceedings of the Association of Asphalt Paving Technologists*, 30, 1961, pp. 188–222.

213. Mun, S., Chehab, G. R., and Kim, Y. R., Determination of time domain viscoelastic functions using optimized interconversion techniques, *Road Materials and Pavement Design*, 8, 2007, pp. 351–365.

214. *NCHRP Design Guide, Mechanistic-empirical design of new & rehabilitated pavement structure*, 1-37A, 2004, http://onlinepubs.trb.org/onlinepubs/archive/mepdg/guide.htm, last accessed November, 2021.

215. Neville, A. M., *Properties of concrete*, Longman, Thomson Press (India) Ltd., 1st Indian Reprint, 2000.

216. Neville, A. M., and Brooks J. J., *Concrete technology*, International Student Edition reprint, Longman Group, 1999.

217. Nunn, M., Brown, A., Weston, D., and Nicholls, J. C., *Design of long-life flexible pavements for heavy traffic*, Report No. 250, Transportation Research Laboratory, Berkshire, United Kingdom, 1997.

218. Onu, G., Finite elements on generalized elastic foundation in Timoshenko beam theory, *Journal of Engineering Mechanics*, 134(9), 2008, pp. 763–776.

219. Ouyang, Y., and Madanat, S., Optimal scheduling of rehabilitation activities for multiple pavement facilities: exact and approximate solutions, *Transportation Research Part A: Policy and Practice*, 38(5), 2004, pp. 347–365

220. Packard, R. G., *Design of concrete airport pavement*, Portland Cement Association, reprint 1995.

221. Papagiannakis, A. T., and Masad, E. A., *Pavement design and materials*, John Wiley & Sons, 2007.

222. Park, S. W., and Kim, Y. R., Analysis of layered viscoelastic system with transient temperature, *Journal of Engineering Mechanics*, 124(2), 1998, pp. 223–231.

223. Park, S. W., and Kim, Y. R., Fitting Prony-series viscoelastic models with power-law presmoothing, *Journal of Materials in Civil Engineering*, 13(1), 2001, pp. 26–32.

224. *Pavement design manual - asphalt pavements and overlays for road traffic*, Shell International Petroleum Company Limited, London, 1978.

225. *Guide to pavement technology part 2: pavement structural design*, AGPT02-12. Austroads Ltd., 2012, Sydney.

226. Peattie, K. R., Stress and strain factors for three layered elastic systems, *Highway Research Bulletin*, 342, Washington, D. C., 1962, pp. 215–253.

227. Pellinan, T. K., and Witczak, M. W., Stress dependent master curve construction for dynamic (complex) modulus, *Proceedings of the Association of Asphalt Paving Technologists*, 71, 2002, pp. 281–309.

228. Peshkin, D. G., Hoerner, T. E., and Zimmerman, K. A., *Optimal timing of pavement preventive maintenance treatment applications*, Report No. 523, NCHRP, TRB, Washington, D. C., 2004.

229. Pister, K. S., Viscoelastic plate on a viscoelastic foundation, *Journal of the Engineering Mechanics Division*, 87 (EM1), 1961, pp. 43–54.

230. Poulos, H. G., and Davis, E. H., *Elastic solutions for soil and rock mechanics*, John Wiley & Sons Inc., 1974.

231. Porter, M. L., Dowel bar optimization: phases I and II, Final Report, Center for Portland Cement Concrete Pavement Technology, Iowa State University, 2001.

232. Pronk, A. C., *Evaluation of dissipated energy for the interpretation of fatigue measurements in crack initiation phase*, Report No. P.DWW95001, Road and Hydraulic Engineering Division, Ministry of Transport, Public Works and Water Management, Delft.

233. Quintus, H. L. V., Mallela, J., Bonaquist, R., Schwartz, C. W., and Carvalho, R. L., *Calibration of rutting models for structural and mix design*, Report No. 719, NCHRP, TRB, Washington, D. C., 2012.

234. Rajbongshi, P., and Das, A., Thermal fatigue considerations in asphalt pavement design, *International Journal of Pavement Research and Technology*, 1(4), 2008, pp. 129–134.

235. Rajbongshi, P., and Das, A., Estimation of temperature stress and low-temperature crack spacing in asphalt pavements, *Journal of Transportation Engineering*, 135(10), 2009, pp. 745–752.

236. Rajbongshi, P., and Das, A., Estimation of structural reliability of asphalt pavement for mixed axle loading conditions, *Proceedings of the 6th Int. Conference of Roads and Airfield Pavement Technology (ICPT)*, Sapporo, Japan, 2008, pp. 35–42.

237. Rajbongshi, P., and Das, A., Optimal asphalt pavement design considering cost and reliability, *Journal of Transportation Engineering*, 134(6), 2008, pp. 255–261.

238. Rajbongshi, P. and Das, A., Temperature stresses in concrete pavement - a review, *International Conference on Civil Engineering in the New Millennium : Opportunities and Challenges (CENeM 2007)*, Bengal Engineering and Science University, Shibpur, January 11–14, 2007, Vol. III, pp. 2080–2090.

239. Ramsamooj, D. V., Stresses in jointed rigid pavement, *Journal of Transportation Engineering*, 125(2), 1999, pp. 101–107.

240. Rao, T. S. C. S., Craus, J., Deacon, J. A., and Monismith, C. L., *Fatigue response of asphalt mixes*, Institute of Transportation Studies, SHRP-A-003-A, University of California Berkeley, California, 1990.

241. Reddy, J. N., *An introduction to the finite element method*, 4th edition, McGraw-Hill, 2020.

242. Reese, R., Properties of aged asphalt binder related to asphalt concrete fatigue life, *Proceedings of the Association of Asphalt Paving Technologists*, 66, 1997, pp. 604–632.

243. Reissner, M. E., On the theory of beams resting on a yielding foundation, *Proceedings of the National Academy of Sciences*, 23(6), 1937, 328–333.

244. Rhines, W. J., Elastic-plastic foundation model, *Journal of Soil Mechanics and Foundation Division*, 95, 1960, pp. 819–826.

245. Richardson, J. M., and Armaghani, J. M., Stress caused by temperature gradient in Portland cement concrete pavements, *Transportation Research Record*, 1121, 1987, pp. 7–13.

246. Roque, R., Zou, J., Kim, Y. R., Baek, C., Thirunavukkarasu, S., Underwood, B. S., and Guddati, M. N., *Top-Down cracking of hot-mix asphalt layers: models for initiation and propagation*, Web-only document No. 162, NCHRP, TRB, Washington, D. C., 2010.

247. Saal, R. N. S., and Pell, P. S., Fatigue of bituminous road mixes, *Kolloid Zeitschrift*, 171(1), 1960, pp. 61–71.

248. Saboo, N., and Das, A., Research trends in materials and design of asphalt pavements, *Transportation Research in India - Practices and Future Directions*, Springer Transactions in Civil and Environmental Engineering, Editors: Maurya, A. K., Vanajakshi, L. D., Arkatkar, A., Sahu, P. 2021 (in press).

249. Sadd, M. H., *Elasticity - theory, applications and numerics*, Academic Press, Elsevier, 2005.

250. Sadd, M. H., Dai, Q., Parameswaran, V., and Shukla, A., Microstructural simulation of asphalt materials: modeling and experimental studies, *Journal of Materials in Civil Engineering*, 16(2), 2004, pp. 107–115.

251. Safaei, F., Lee, J., Nascimento, L. A. H. do, Hintz, C., and Kim, Y. R., Implications of warm-mix asphalt on long-term oxidative ageing and fatigue performance of asphalt binders and mixtures, *Road Materials and Pavement Design*, 15(sup1), 2014, pp. 45–61.

252. Salamaa, H. K., and Chatti, K., Evaluation of fatigue and rut damage prediction methods for asphalt concrete pavements subjected to multiple axle loads, *International Journal of Pavement Engineering*, 12(1), 2011, pp. 25–36.

253. Sanchez-Silva, M., Arroyo, O., Junca, M., Caro, S., and Caicedo, B., Reliability based design optimization of asphalt pavements, *International Journal of Pavement Engineering*, 6(4), 2005, pp. 281–294.

254. Sarkar, A., Numerical comparison of flexible pavement dynamic response under different axles, *International Journal of Pavement Engineering*, 17(5), 2016, pp. 377–387.

255. Sawant, V., Dynamic analysis of rigid pavement with vehicle–pavement interaction, *International Journal of Pavement Engineering*, 10(1), 2009, pp. 63–72.

256. Schapery, R. A., and Park, S. W., Methods of interconversion between linear viscoelastic material functions. Part II — an approximate analytical method, *International Journal of Solids and Structures*, 36(11), 1999, pp. 1677–1699.

257. Schiffman, R. L., General analysis of stresses and displacements in layered elastic systems, *Proceedings of the 1st International Conference on Structural Design of Asphalt Pavements*, University of Michigan, Ann Arbor, 1962, pp. 365–375.

258. Selvadurai, A. P. S., *Elastic analysis of soil-foundation interaction*, Development of Geotechnical Engineering, Elsevier Scientific Publishing Company, Vol. 17, 1979.

259. Selvadurai, A. P. S., On Boussinesq's problem, *International Journal of Engineering Science*, 39, 2001, pp. 317–322.

260. Setiadji, B. H., and Fwa, T. F., Examining $k - E$ relationship of pavement subgrade based on load-deflection consideration, *Journal of Transportation Engineering*, 135(3), 2009, pp. 140–148.

261. Seyhan, U., and Tutumluer, E., Anisotropic modular ratios as unbound aggregate performance indicators, *Journal of Materials in Civil Engineering*, 14(5), 2002, pp. 409–416.

262. Shakiba, M., Gamez, A., Al-Qadi, I. L., and Little, D. N., Introducing realistic tire–pavement contact stresses into Pavement Analysis using Nonlinear Damage Approach (PANDA), *International Journal of Pavement Engineering*, 18(11), 2017, pp. 1027–1038.

263. Shakiba, M., Al-Rub, R. K. A., Darabi, M. K., You, T., Masad, E. A., and Little, D. N., Continuum coupled moisture–mechanical damage model for asphalt concrete, Transportation Research Record, 2372, 2013, pp. 72–82.

264. Shen, W., and Kirkner, D. J., Distributed thermal cracking of AC pavement with frictional constraint, *Journal of Engineering Mechanics*, 125(5), 1999, pp. 554–560.

265. Shen, W., and Kirkner, D. J., Thermal cracking of viscoelastic asphalt-concrete pavement, *Journal of Engineering Mechanics*, 127(7), 2001, pp. 700–709.

266. Shi, X. P., Fwa, T. F., and Tan, S. A., Warping stresses in concrete pavements on Pasternak foundation, *Journal of Transportation Engineering*, 119(6), 1993, pp. 905–913.

267. Shi, X. P., Tan, S. A., and Fwa, T. F., Rectangular thick plate with free edges on Pasternak foundation, *Journal of Transportation Engineering*, 120(5), 1994, pp. 971–988.

268. Shook, J. F., and Finn, F. N., Thickness design relations for asphalt pavements, *Proceedings of the 1st International Conference on Structural Design of Flexible Pavements*, University of Michigan, Ann Arbor, Michigan, 1962, pp. 640–687.

269. Shook, J. F., Finn, F. N., Witczak, M. W., and Monisminth, C. L., Thickness design of asphalt pavements, The Asphalt Institute Method, *Proceedings of the 5th International Conference on Structural Design of Flexible Pavement*, Delft University of Technology, Delft, 1982, pp. 17–44.

270. Shukla, P. K., and Das, A., A re-visit to the development of fatigue and rutting equations used for asphalt pavement design, *International Journal of Pavement Engineering*, 9(5), 2008, pp. 355–364.

271. Singh, S. J., Static deformation of a transversely isotropic multilayered half-space by surface loads, *Physics of the Earth and Planetary Interiors*, 42, 1986, pp. 263–273.

272. Smallridge, M., and Jacob, A., *The ASCE Port and Intermodal Yard Pavement Design Guide*, Ports'01, Norfolk, 2001, p.10.

273. Chapter-10, Pavement design, Manuals and Policies, South African National Road Agency, SANRAL, https://www.nra.co.za/manuals-policies-technical-specifications/manuals-policies/, last accessed November, 2021.

274. Sousa, J. B., Deacon, J. A., Weissman, S., Harvey, J. T., Monismith, C. L., Leahy, R. B., Paulsen, G., and Coplantz, J. S., *Permanent deformation response of asphalt-aggregate mixes*, Report no. SHRP-A-415, Strategic Highway Research Program, Washington, D.C., 1994.

275. Stubstad, R.N., Tayabji, S. D., and Lukanen, E. O., *LTPP data analysis: variation in pavement design inputs*, Final Report, Web Document No. 48, NCHRP, TRB, Washington, D.C., 2002, http://gulliver.trb.org/publications/nchrp/nchrp_w48.pdf. Last accessed on November, 2021.

276. Sun, L., Hudson, W. R., and Zhang, Z., Empirical-mechanistic method based stochastic modeling of fatigue damage to predict flexible pavement cracking for transportation infrastructure management, *Journal of Transportation Engineering*, 129(2), 2003, pp. 109–117.

277. Sun, L., A closed-form solution of beam on viscoelastic subgrade subjected to moving loads, *Computers and Structures*, 80, 2002, pp. 1–8.

278. Svasdisant, T., Schorsch, M., Baladi, G. Y., and Pinyosunun, S., Mechanistic analysis of top-down cracks in asphalt pavements, *Transportation Research Record*, 1809, 2002, pp. 126–135.

279. Tabatabaie, A. M., and Barenberg, E. J., Structural analysis of concrete pavement systems, *Journal of Transportation Engineering*, 106(5), 1980, pp. 493–506.

280. Tang, T., Zollinger, G., and Senadheera, S., Analysis of concave curling in concrete slabs, *Journal of Transportation Engineering*, 119(4), 1993, pp. 618–633.

281. Tayebali, A. A., Deacon, J. A., Coplantz, J. S., Harvey, J. T., and Monismith, C. L., *Fatigue Response of asphalt-aggregate mixes*, SHRP-A-404, Institute of Transportation Studies, University of California, Berkeley, 1994.

282. Terzaghi, K., Evaluation of coefficients of subgrade reaction, *Géotechnique*, 5(4), 1995, pp. 41–50.

283. Timm, D. H., Guzina, B. B., and Voller, V. R., Prediction of thermal crack spacing. *International Journal of Solids and Structures*, 40, 2003, 125–142.

284. Timm, D. H., Newcomb, D. E., Briggison, B., and Galambos, T. V., *Incorporation of reliability into the Minnesota mechanistic-empirical pavement design method*, Final Report, Submitted to Minnesota Department of Transportation, Department of Civil Engineering, Minnesota University, Minneapolis, 1999.

285. Timm, D. H., Newcomb, D. E., and Galambos, T. V., *Incorporation of reliability into mechanistic-empirical pavement design*, Transportation Research Record, 1730, 2000, pp. 73–80.

286. Timoshenko, S. P., and Woinowky-Krieger, S., *Theory of plates and shells*, McGraw Hill, New York, 1959.

287. Timoshenko, S. P., and Gere, J. M. *Theory of Elastic Stability*, International Student Edition (2nd edition), McGraw-Hill, 1985.

288. Timoshenko, S., P. and Goodier, J. N., *Theory of Elasticity*, McGraw-Hill, 1934.

289. Titi, H. H., Elias, M. B., and Helwany, S., *Determination of typical resilient modulus values for selected soils in Wisconsin*, University of Wisconsin - Milwaukee, Submitted to The Wisconsin Department of Transportation, 2006.

290. *The handbook of highway engineering*, Edited by: Fwa, T. F., CRC Press, Taylor & Francis Group, 2006.

291. Theyse, H. L., de Beer, M., and Rust, F. C., Overview of the South African mechanistic pavement design method, *Transportation Research Record*, 1539, 1996, pp. 6–17.

292. *Thickness design for concrete highway and street pavements*, Portland Cement Association, 1984, Skokie.

293. Thompson, M. R., Barenberg, E., Brown, S. F., Darter, M. M., Larson, G., Witczak, M., and El-Basyouny, M., *Independent review of the Mechanistic-Empirical pavement design guide and software*, Research results digest No. 307, NCHRP, TRB, Washington, D.C., 2006. http://onlinepubs.trb.org/onlinepubs/nchrp/nchrp_rrd_307.pdf, last accessed November, 2021.

294. Thompson, M. R., and Elliott, R. P., ILLI-PAVE based response algorithms for design of conventional flexible pavements, *Transportation Research Record*, 1043, 1985, pp. 50–57.

295. *Thickness design – asphalt pavements for highways and streets*, Asphalt Institute, Manual Series No. 1 (MS-1), 1991.

296. *Thickness design of concrete pavements*, Portland Cement Association publication, ISO10P, 1974.

297. Tseng, K. H., and Lytton, R. L., Fatigue damage properties of asphaltic pavements, *Transportation Research Record*, 1286, 1990, pp. 150–163.

298. Tsunokawa, K., and Schofer, J. L., Trend curve optimal control model for highway pavement maintenance: case study and evaluation, *Transportation Research*, 28A, 1994, pp. 151–166.

299. Tutumluer, E., *Practices for unbound aggregate pavement Layers - a synthesis of highway practice*, Synthesis 445, NCHRP, TRB, Washington, D. C., 2013.

300. Ullidtz, P., *Pavement analysis*, 19, Elsevier, 1986.

301. Underwood, B. S., and Kim, Y. R., Determination of the appropriate representative elastic modulus for asphalt concrete, *International Journal of Pavement Engineering*, 10(2), 2009, pp. 77–86.

302. Underwood S., Heidari, A. H., Guddati, M., and Kim, Y. R., Experimental investigation of anisotropy in asphalt concrete, *Transportation Research Record*, 1929, 2005, pp. 238–247.

303. Uzan, J., Characterization of granular materials, *Transportation Research Record*, 1022, 1985, pp. 52–59.

304. Vallabhan, C. V. G., and Das, Y. C., Modified Vlasov model for beams on elastic foundations, *Journal of Geotechnical Engineering*, 117(6), 1991, pp. 956–966.

305. Van Dijk, W. Practical fatigue characterization of bituminous mixes, *Proceedings of the Association of Asphalt Paving Technologists*, 44, 1975, pp. 38–74.

306. Van Dijk, W., Moreaud, H., Quedeville, A., and Ugé, P., The fatigue of bitumen and bituminous mixes, *Proceedings of the 3rd International Conference on the Structural Design of Asphalt Pavements*, Vol. 1, London, 1972, pp. 354–366.

307. Van Dijk, W., and Visser, W., The energy approach to fatigue for pavement design, *Proceedings of the Association of Asphalt Paving Technologists*, 46, 1977, pp. 1–40.

308. Varma, R. K., Padmarekha, A., Ravindran, P., Bahia, H. U., and Krishnan, J. M., Evolution of energy dissipation during four-point bending of bituminous mixtures, *Road Materials and Pavement Design*, 18(sup2), 2017, pp. 252–263.

309. Ventsel, E., and Krauthammer, T., *Thin plates and shells - theory, analysis and application*, Marcel Dekker, 2001.

310. Verstraeten, J., Stresses and displacements in elastic layered systems, general theory – numerical stress calculation in four-layered systems with continuous interfaces, *Proceeding of 2nd International Conference of Structural Design of Asphalt Pavements*, Ann Arbor, 1967, pp. 277–290.

311. Verstraeten, J., Moduli and critical strains in repeated bending of bituminous mixes application to pavement design, *Proceeding of 3rd International Conference of Structural Design of Asphalt Pavements*, London, Vol. 1, 1972, pp. 729–738.

312. Vinson, T. S., Janoo, V. C., and Haas, R. C. G., *Summary report - low temperature and thermal fatigue cracking*, Report SHRP-A/IR-90-001, Strategic Highway Research Program, Washington, D. C., 1990.

313. Vlasov, V. Z., and Leont'ev, M. N., *Beams, plates and shells on elastic foundations*, Israel Program for Scientific Translations, translated from Russian, 1960.

314. Von Quintus, H. L., *Hot mix asphalt layer thickness design for longer life bituminous pavements*, Transportation Research Circular, No. 503, TRB, Washington, D. C., 2001, pp. 66–78.

315. Wang, D., *Analytical solutions for temperature profile prediction in multi-layered pavement systems*, Ph.D. Dissertation, Dept. of Civil and Environmental Engineering, Univ. of Illinois at Urbana- Champaign, IL, Open access http://hdl.handle.net/2142/18241, last accessed November, 2021.

316. Wang, D., Analytical approach to predict temperature profile in a multilayered pavement system based on measured surface temperature data, *Journal of Transportation Engineering*, 138(5), 2012, pp. 674–679.

317. Westergaard, H. M., Stresses in concrete pavements computed by theoretical analysis, *Public Roads*, 7, 1926, pp. 25–35.

318. Westergaard, H. M., Analysis of stresses in concrete pavement due to variations in temperature, *Highway Research Record*, 6, Washington, D.C., 1927, pp. 201–217.

319. Westergaard, H. M., Stresses in concrete runways of airports, *Proceedings of 19th Annual Meeting*, Highway Research Board, Washington, D.C., 1939, pp. 197–203.

320. Westergaard, H. M., New formulas for stresses in concrete pavements of airfields, *Proceedings of ASCE*, 73, 1947, pp. 687–701.

321. White, T. D., Haddock, J. E., Hand, A. J. T., and Fang, H., *Contributions of pavement structural layers to rutting of hot mix asphalt pavements*, Report No. 468, NCHRP, TRB, Washington, D. C., 2002.

322. Williams, M. L., Landel, R. F., and Ferry, J. D., The temperature dependence of relaxation mechanisms in polymers and other glassforming liquids, *Journal of American Chemical Society*, 77, 1955, pp. 3701–3707.

323. Witczak, M., Mamlouk, M., Souliman, M., Zeiada, W., *Laboratory validation of an endurance limit for asphalt pavements*, Report No. 762, NCHRP, TRB, Washington, D. C., 2013.

324. Yan, K., Shi, T., You, L., and Man, J., Spectral element method for dynamic response of multilayered half medium subjected to harmonic moving Load, *International Journal of Geomechanics*, 2018, 18(12), 04018161.

325. Yang, N. C., *Design of functional pavements*, McGraw-Hill Book Company, 1972.

326. Yavuzturk, C., K. K. C. A., Assessment of temperature fluctuations in asphalt pavements due to thermal environmental conditions using a two-dimensional, transient finite difference approach, *Journal of Materials in Civil Engineering*, 17(4), 2005, pp. 465–475.

327. Yin, H., An analytical procedure for strain response prediction of flexible pavement, *International Journal of Pavement Engineering*, 14(5), 2013, pp. 486–497.

328. Yoder, E. J., and Witczak, M. W., *Principles of pavement design*, 2nd edition, John Wiley & Sons, Inc., 1975.

329. Yoo, P. J., Al-Qadi, I. L., Elseifi, M. A., and Janajreh, I. Flexible pavement responses to different loading amplitudes considering layer interface condition and lateral shear forces, *International Journal of Pavement Engineering*, 7(1), 2006, pp. 73–86.

330. You, Z., and Buttlar, W. G., Discrete element modeling to predict the modulus of asphalt concrete mixtures, *Journal of Materials in Civil Engineering*, 16(2), 2004, pp. 140–146.

331. Yusoff, M. I. Md., Chailleux, E., and Airey, G. D., A comparative study of the influence of shift factor equations on master curve construction, *International Journal of Pavement Research and Technology*, 4(6), 2011, pp. 324–336.

332. Yusoff, M. I. Md., Shaw, M. T., Airey, G. D., Modelling the linear viscoelastic rheological properties of bituminous binders, *Construction and Building Materials*, 25, 2011, pp. 2171–2189.

333. Zeinkiewicz, O. C., Valliappan, S., and King, I. P., Stress analysis of rock as 'no-tension' material, *Géotechnique*, 18, 1968, pp. 56–66.

334. Zejun, H., Jin, Z., Luanluan, X., and Hongyuan, F., and Ziwei, X., Dynamic simulation of FWD tests on flexible transversely isotropic pavements with imperfect interfaces, *Computers and Geotechnics*, 130, 2021, 103914.

335. Zhang, H., Anupam, K., Scarpas, T., Kasbergen, C., Erkens, S., and Khateeb, L. A., Continuum-based micromechanical models for asphalt materials: current practices & beyond, *Construction and Building Materials*, 260(10), 2020, 119675.

336. Zhang, J., Fwa, T. F., Tan, K. H., and Shi, X. P., Five-slab thick-plate model for concrete pavement, *Road Materials and Pavement Design*, 2000, 1, pp. 10–34.

337. Zhang, J., Fwa, T. F., Tan, K. H., and Shi, X. P., Model for nonlinear thermal effect on pavement warping stresses, *Journal of Transportation Engineering*, 129(6), 2003, pp. 695–702.

338. Zhang, J., and Li, V. C., Influence of supporting base characteristics on shrinkage-induced stresses in concrete pavements, *Journal of Transportation Engineering*, 127(6), 2001, pp. 455–462.

339. Zhang, Y., Gu, F., Luo, X., Birgisson, B., and Lytton, R. L., Modeling Stress-Dependent anisotropic elastoplastic unbound granular base in flexible pavements, *Transportation Research Record*, 2672(52), 2018, pp. 46–56.

340. Zhao, Y., Ni, Y., and Zeng, W., A consistent approach for characterising asphalt concrete based on generalised Maxwell or Kelvin model, *Road Materials and Pavement Design*, 15(3), 2014, pp. 674–690.

341. Zhao, Y. Zhou, C., Zeng, W., and Ni, Y., Accurate determination of near-surface responses of asphalt pavements, *Road Materials and Pavement Design*, 16(1), 2015, pp. 186–199

342. Zhu, H., and Sun, L., Mechanistic rutting prediction using a two-stage viscoelastic-viscoplastic damage constitutive model of asphalt mixtures, *Journal of Engineering Mechanics*, 139(11), 2013, pp. 1577–1591.

343. Zubeck, H. K., and Vinson, T. S., Prediction of low-temperature cracking of asphalt concrete mixtures with thermal stress restrained specimen test results, *Transportation Research Record*, 1545, 1996, pp. 50–58.

Index

For Product Safety Concerns and Information please contact our EU representative GPSR@taylorandfrancis.com
Taylor & Francis Verlag GmbH, Kaufingerstraße 24, 80331 München, Germany

www.ingramcontent.com/pod-product-compliance
Lightning Source LLC
Chambersburg PA
CBHW070724220326
41598CB00024BA/3285
9781032041575